C000132698

Dear Zoe —

lots of love — Miles X

A Journey through Cosmic Consciousness

Wes Jamroz

Troubadour Publications

A Journey through Cosmic Consciousness

Editing: *Dominique Hugon, Patrick Barnard*
Cover illustration: *Sandra Viscuso* (sandraviscuso.com)

Copyright © 2019 by Troubadour Publications. All rights reserved.

No part of this work may be reproduced or transmitted in any form or by any means, electronic or mechanical, including photocopying and recording, or by any information storage or retrieval system without the prior written permission of Troubadour Publications.

Montreal, QC, Canada

TroubadourPubs@aol.com
http://www.troubadourpublications

ISBN: 978-1-928060-11-6

A Note on quoted Texts

The Biblical quotes are from *The Holy Bible* (Chancellor Press, London, 1981).

The quotes from *The Gospel of Thomas* are from a revised and updated edition by Marvin Meyer (Harper One, New York, 1992).

The Koranic quotes are from *The Holy Qur'an,* translated by Abdullah Yusuf Ali (Wordsworth Editions Limited, Ware, Hertfordshire, 2000).

The extracts from Rumi's *Mathnawi* are adapted by the author from Reynold A. Nicholson's translation *The Mathnawi of Jalaluddin Rumi* (E.J. W. Gibb Memorial, Cambridge, 1960).

The quotations from Shakespeare's plays and sonnets are from *The Riverside Shakespeare-Second Edition* (Houghton Mifflin Company, Boston, 1997). Punctuation in the quotes from the plays was adjusted according to the original version of *The First Folio of Shakespeare* (W.W. Norton & Company, New York, 1996).

Table of Contents

Observer, Observed, and the Act of Observation

> There are two modes of knowledge, through argument and experience. Argument brings conclusions and compels us to concede them, but it does not cause certainty nor remove doubts in order that the mind may remain at rest in truth, unless this is provided by experience.
> (*Roger Bacon*)

In recent years the subject of consciousness has become part of science. It has been suggested that consciousness is more fundamental than any known physical phenomenon.

Progress in the understanding of consciousness has been made through the development of quantum theory. In this theory, the nature of consciousness is treated as a field phenomenon, an analogue to the quantum field. Consequently, this field is characterized by the generalized principles described by quantum physics. These principles indicate that everything in the physical world is somewhat connected. The interconnectedness of everything is clearly demonstrated in the non-local interactions of the quantum universe, where objects seem to instantaneously know about each other's state, even when separated by large distances.

Quantum theory implies that objects are superpositions of entities whose underlying structure is wave-like. Therefore, they may be represented by a wavefunction. The wavefunction of a system collapses under observation, i.e., a wave is transformed into an object. Consequently, quantum theory assigns a fundamental role to the act of observation. This theory has opened the door to a new vision of the Universe where observer, observed, and the act of observation are interlocked.[1] The fact that the wavefunction of a system collapses under observation indicates that there is an interaction between consciousness and matter. It is this feature that led to a belief that consciousness and matter are two complementary aspects of one reality.

This would mean that consciousness constitutes a primordial quantum field that consists of, among other things, a set of ideas and concepts. These ideas and concepts are capable of turning into substance. Substance is manifested as the Universe and the human race. Humans, through their efforts, are capable of performing the function of a conscious observer. Their act of observation results in the collapse and localization of the field of consciousness. The localization leads to the appearance of matter. Mankind is the highest product of that matter. Humans have self-awareness to contemplate their own origins. This contemplation leads to a growth of consciousness. In this way, the loop of this self-consistency *seems* to be closed.

Furthermore, it is assumed that consciousness is primary in the Universe. It is considered to be the source of reality; a reality that is self-efficient, capable of engendering complex systems on the micro and macro scale, self-regulating and holistic. Nothing can exist outside its influence. And, ultimately, it must also be present within the Universe. It is with respect to this aspect that, according to science, consciousness' role is different from that of an overseer, i.e., the role assigned by conventional religions to God or gods.[2]

---

[1] "How Consciousness Becomes the Physical Universe," M. Kafos e*t al.*, (included in *Consciousness Became the Universe, 2nd edition*, Science Publishers, Middletown, DE, 2017, p. 6).

[2] *Ibid.*

On the other hand, our sense of reality is different from its mathematical models as proposed by quantum theory. Instead, our daily experiences are perceived as local and classical. It has been suggested, therefore, that this difference in the nature of perceived reality is due to the principle of veiled non-locality.[3] Veiled non-locality allows consciousness to operate and present what we experience as objective and local reality. The veiling of reality is in consonance with the idea of the mind constructing its reality. Such a veiling also occurs in the scientific process which filters out and discards a huge portion of human experience – almost everything one would classify as subjective. As far as the brain is concerned, neural filtering is taking place in all models, whether they are scientific, artistic, psychotic, or religious. The brain is a processor of inputs, not a mirror of reality. The brain operates in time and space, having linear thoughts that are the end point of a selective filtering process. So whatever is outside time and space is inconceivable, and unfiltered reality would probably blow the brain's circuits, or simply be blanked out.[4]

Despite this explanation, the issues related to consciousness still remain the most difficult problem for science. There are still many fundamental questions that need to be addressed and answers to them seem to be beyond the grasp of known scientific models. It is hoped, however, that the combined efforts of quantum physics, biology, cosmology and neuroscience will allow to determine in scientific terms what role consciousness plays in the Universe.

There is another approach that may greatly accelerate the development of our understanding of consciousness and its role in the Universe and in human life. First, however, let's examine the framework within which modern science operates.

---

[3] "Perceived Reality, Quantum Mechanics, and Consciousness," S. Kak *et al.*, (included in *Consciousness Became the Universe, 2nd edition*, Science Publishers, Middletown, DE, 2017, p. 16).

[4] *Ibid.*

Contrary to a widely accepted view, science is based on belief. The main driving force of science is the belief that there is no need for outside intervention. Consequently, scientific efforts are now focussed on a theory in which the Universe would be completely self-contained, having no boundaries or edge, and would have neither beginning nor end. There is no need for any sort of creator. In other words, science aims at un-creating the source of creation.

Science believes that the Universe and human life may be explained by a single theory based on advanced mathematics. Furthermore, it is believed that the ultimate mathematical model has to be aesthetically satisfying. This is why scientific models and theories are initially put forward for aesthetic reasons. It is quite interesting that an aspect of beauty plays such an important role in science. Somehow, the simplicity of mathematical formulas which contain an incredible intellectual richness is strongly attractive – just like a piece of profound art. Formulated in the 19th century, Maxwell's equations may serve as a good example of such an art:

$$\Box E = 0$$
$$\Box B = 0$$

These few doodle-like symbols contain the entire knowledge of classical electro-magnetism. Neither a painting nor a sculpture is capable to express the depth of intellectual richness that is contained in these few graphical symbols. Of course, one would have to learn the meaning of these symbols and be familiar with the rules of their operation to fully appreciate the "secret" knowledge that is hidden in them. For a person who is not familiar with this specific art, these doodles have neither meaning nor importance.

Before suggesting an aesthetically pleasing model, scientists make a series of assumptions and approximations. These assumptions and approximations are then adjusted to fit into a set of available experimental data. Usually, these models last for a while, till the development of more and more precise instrumentation leads to new data. Then, the models are corrected, adjusted, some new constants are introduced and a new model is worked out. And so on and so forth.

It is this belief in aesthetically pleasing forms that constitutes the modus operandi of the entire scientific enterprise. Sometimes, like religious sectarians, scientists are tempted to manipulate their own doctrine. There is an interesting episode that nicely illustrates such a tendency. One of the outcomes of the Big Bang theory was the idea that time has a beginning. This concept seemed to indicate some sort of divine intervention. When the Catholic Church seized on the Big Bang and in 1951 officially pronounced it to be in accordance with the Bible, a group of leading physicists panicked.[5] This was like a threatening lightning from the sky! Afterwards, a number of attempts were initiated to avoid the conclusion that there had been any sort of a big bang. The alternative model that gained the widest support within the scientific community was called the steady state theory. This theory was based on the idea that the Universe looks roughly the same at all times as well as at all points of space. However, later observations contradicted this idea and the steady state theory was abandoned.

Scientists like to compare the process of creation of life to a situation where a horde of monkeys hammers away on typewriters over a long length of time. Most of what they write will be garbage, but by pure chance one of them will type out one of Shakespeare's sonnets. Such a sonnet would be equivalent to the development within one of the galaxies of a planet where the conditions would be right for the appearance of complicated self-replicating organisms, like ourselves, who would be capable of asking the question: why are we here? This often-used example with poor monkeys misses one important point. As there is no logical link

---

[5] *A Brief History of Time*, Stephen Hawking (Bantam Books, New York, 1988, p. 46).

between alphabetical permutations and the sublime meaning of the sonnets, how would such a sonnet know that it is itself a sonnet?

The model that is presently proposed by science has yet to provide satisfactory answers to the questions: Has man, by some accident, been overdesigned and equipped with faculties that he does not really need to survive? Or, is this an intentional hint indicating that humans are capable of developing different kinds of faculties which are needed to fulfil their cosmic function?

The scientific approach has to be based on known and verifiable facts, repeatable experiments, and on a preference for data over opinion. To be approved, the data have to be confirmed by other experiments carried out in the same conditions. The results have to be reproduced regardless of the state of mind of the experimenter. In other words, an experiment has to provide exactly the same results, regardless of whether a wise man or a fool performs it.

This approach excludes from its consideration any data or information obtained by certain individuals who access heightened states of consciousness. It is through these heightened states that it is possible to overcome the veiling of the mind and break through the limitations of space and time. These sorts of data are classified as "mystical" experiences and, by definition, do not constitute reliable information. In popular understanding, mysticism is belief-based. And as such, is not scientifically valid. Yet, it is mysticism that offers that other approach that may greatly accelerate the development of our understanding of consciousness and its role in the Universe and in human life.

Contrary to popular opinion, mystics do not follow their beliefs. They do not believe because their knowledge has been gained from experience. In this context, the term "mystics" is not quite an adequate term. In reality, mystics should be referred to as "craftsmen," whose craft and expertise are in the field of consciousness. They know and apply laws that rule over the

Universe and life. Therefore, they are more like technicians and engineers, whose job is to preserve an *environment* that they are in charge of. While science is still discovering and learning, the mystics have already acquired a complete knowledge. From time to time, they have to update their methods and techniques in accordance with the time and places in which they live.

There is another difference between scientific and mystical approaches. Namely, in the scientific approach, the conscious state of the observer is necessarily exclusive. In the case of mysticism, only an observer whose mind has been freed from worldly attachments is capable of providing adequate data. And it is this requirement that has put a barrier separating science from access to meaningful mystical data. In this way, the scientific exploration of the Universe has been cut-off from access to a substantial pool of information. Now, however, this situation may change as a result of the development of quantum theory. Quantum theory has introduced an entirely new set of terms and concepts which seem to echo various mystical experiences. Therefore, now may be the right time to revisit the pool of mystical data and express them in the new language of physics and modern cosmology.

Mystical data have been gathered over a period of several thousands of years. This type of information, if accepted, would provide several critical hints that could greatly accelerate the efficiency of present and future exploration of the Universe. The underlying experience of the mystics indicates that there are layers of the human mind that are capable of accessing heightened levels of consciousness. While science has limited itself to the lowest level of man's consciousness, these higher levels of consciousness open up possibilities of knowledge-yielding experiences. The difficulty with some of the mystical experiences is that they are described in elliptical and symbolic language, either for lack of adequate terminology, as a safety measure, or to prevent them from abuse by the ignorant and greedy. The mystical data may take the form of a scripture or a philosophical treatise, poetry or tales, architectural structures or a garden. These forms enable people to absorb ideas which the ordinary patterns of their thinking would prevent them from digesting. These forms have been used to present a picture of reality more in harmony with people's inner needs than is possible by means of intellectual discourses. All these forms are expressions of an experienced reality rather than intellectually and aesthetically satisfying explanations. This is why they may seem to be contradictory or exaggerations. However, when these experiences are aligned in accordance with the perceptive ability of the audience, they become quite consistent and provide meaningful insights into the deeper structure of the human mind.

Sir Roger Penrose, an English mathematical physicist, concluded his bestselling book, *The Emperor's New Mind*, with -what he called- "a child's view."[6] This view is a set of basic questions which children are not afraid to ask while adults are embarrassed to do so. These are questions such as: "What happens to each of our stream of consciousness after we die; where was it before each was born; might we become, or have been, someone else; why do we perceive at all; why are we here; why is there a universe here at all in which we can actually be?" At the end of his 466 page book about science, consciousness, artificial intelligence, quantum physics, and cosmology, Penrose sincerely admits that he is not able to come up with convincing answers to any of those "childish" questions.

A Journey through Cosmic Consciousness is an attempt at providing some background information that would allow addressing these "childish" questions. The provided information is based on mystics' experiences but is expressed in the language of quantum physics and modern cosmology.

[6] *The Emperor's New Mind*, Roger Penrose (Oxford University Press, New York, 1989, p. 447).

The Essence of Beauty

> Oh how much more doth beauty beauteous seem,
> By that sweet ornament which truth doth give …
> (*William Shakespeare*)

A male bowerbird is considered to be among the most behaviorally complex species of birds. It was thought that he belongs to the family of bird-of-paradise. The male bowerbird is a medium-sized bird, up to 25 cm long, with flame orange and golden yellow plumage, elongated neck plumes and yellow-tipped black tail. This bird is named for the extraordinary structure of his bowers. We may call him *Architect*, the bird.

Here are excerpts from an article dedicated to the spectacular beauty and complexity of the bird's bowers. "How Beauty is Making Scientists Rethink Evolution" was published by Ferris Jabr in *The New York Times Magazine*:

> A male flame bowerbird is a creature of incandescent beauty. The hue of his plumage transitions seamlessly from molten red to sunshine yellow. But that radiance is not enough to attract a mate. When males of most bowerbird species are ready to begin courting, they set about building the structure for which they are named: an assemblage of twigs shaped into a spire, corridor or hut. They decorate their bowers with scores of

colorful objects, like flowers, berries, snail shells or, if they are near an urban area, bottle caps and plastic cutlery. Some bowerbirds even arrange the items in their collection from smallest to largest, forming a walkway that makes themselves and their trinkets all the more striking to a female - an optical illusion known as forced perspective that humans did not perfect until the 15th century.[7] …

The bowerbird defies traditional assumptions about animal behavior. Here is a creature that spends hours meticulously curating a cabinet of wonder, grouping his treasures by color and likeness. Here is a creature that single-beakedly builds something far more sophisticated than many celebrated examples of animal toolmaking; the stripped twigs that chimpanzees use to fish termites from their mounds pale in comparison. The bowerbird's bower, as at least one scientist has argued, is nothing less than art. When you consider every element of his courtship - the costumes, dance and sculpture - it evokes a concept beloved by the German composer Richard Wagner: *Gesamtkunstwerk*, a total work of art, one that blends many different forms and stimulates all the senses.

This extravagance is also an affront to the rules of natural selection. Adaptations are meant to be useful - that's the whole point - and the most successful creatures should be the ones best adapted to their particular environments. So what is the evolutionary justification for the bowerbird's ostentatious display? Not only do the bowerbird's colorful feathers and

---

[7] *Note by the author*: This style of false perspective became known in architecture as Mannerism. Mannerism was devised to demonstrate the imperfections of our physical senses. For example, this is done by designing a street with an artificial perspective that would make it to appear wider and longer; or by using plaster as an imitation of marbles. This style of architecture came into full prominence thanks to Gulio Romano, an Italian mannerist architect, painter, and sculptor. Gulio Romano became known in England thanks to Shakespeare (See V.2 of *The Winter's Tale*, when a gentleman discusses how Romano is able to "beguile Nature of her custom, so perfectly he is her ape.")

> elaborate constructions lack obvious value outside courtship, but they also hinder his survival and general well-being, draining precious calories and making him much more noticeable to predators....
>
> Philosophers, scientists and writers have tried to define the essence of beauty for thousands of years. The plurality of their efforts illustrates the immense difficulty of this task. Beauty, they have said, is harmony, goodness, a manifestation of divine perfection, a type of pleasure; that which causes love and longing. ...[8]

It looks like the animal kingdom is driven not just by natural selection. In addition to survival, there is another drive that is equally strong and important. This other drive is ... beauty. Animals, just like scientists, simply find certain aesthetic features to be appealing. What is the origin and purpose of such a profound inner structure of the animal's brain? What is the reason of this enigma of beauty? Could it be that something went wrong with the creative process that it ended up with features that clearly indicate gross overdesign?

Or maybe *beauty* is a form of the act of observation between observer and observed?

To fully understand this, we must go beyond the beginning of everything.

---

[8] "How Beauty is Making Scientists Rethink Evolution," Ferris Jabr (*The New York Times Magazine*, January 9th, 2019).

A Hidden Treasure

> The purified heart is a treasure of Divine light, though its talisman is of earth.
> *(Jalaluddin Rumi)*

Before the beginning, the Absolute was entirely immersed in self-contemplation. There was nothing except the One. This state of singleness encompassed infinite qualities of beauty and perfection. This was like an incredibly precious but undiscovered treasure. The Absolute was unknown.

At one point, the Absolute conceived a longing "to be known." It desired to have a witness of this immaculate beauty and unperturbed perfection. For the treasure to be fully appreciated, it would have to be displayed against an utterly inferior background. This is reflected in the saying, "It is part of the perfection of being that there is imperfection in it." Such a background would provide a contrast, so the qualities of beauty and perfection could be manifested in their full splendour of multiplicity and variety. Then, a perfect witness would be required who would be capable to comprehend and appreciate the grandeur of such an experience. Only the Absolute itself could fully discharge the function of such a perfect witness. Through witnessing itself outside of itself, the process of the synthesis of self-realization would be completed. It was at that point that the idea of creation was conceived. This concept of creation is captured in one of the most cited sayings of Mohammed:

I was a hidden treasure,
and I wished to be known,
So I created a creation.

This statement encompasses the entire purpose of creation and provides us with hints that, if grasped, could outline a direction for current and future scientific exploration of the Universe.

It was then that the Absolute envisaged the Universe with its galaxies, suns, and planets as the needed background for the display of the treasure. Within the Universe, the Earth would be chosen to provide an environment for the accommodation of a vehicle, within which the Absolute would place itself. In other words, the Absolute would descend all the way to the lowest levels of the creation and leave a sample of itself there. This unique vehicle was formed in the shape of mankind. This is how man, a handful of dust, was to become a host of this most rare experience. Afterwards, man would be tasked with an incredibly difficult undertaking. He would have to recognize the overall purpose of his being and then fulfil it accordingly. In other words, man was to share with the Absolute in the realization of the wish, "I wished to be known." Only then, the loop of self-realization would be completed.

It is quite astonishing that the physicists, driven by the beauty and aesthetics of mathematical equations, have been able to envisage that "observer, observed, and the act of observation are interlocked," i.e., to discover the principle of creation and the role of the observer.

The singleness of the Absolute was the state before the process of creation was initiated. Creation is the process of bringing preexisting ideas into existence. This may be compared to bringing into existence structures which were first conceived in the form of designs or drawings.

As the process has been directed from outside of the physical state, it is impossible to describe it by using mathematical equations. Therefore, a different approach is needed to transmit this type of knowledge. Namely, some aspects of the process and man's role in it have been illustrated by using symbols and allegories.

The mystical symbols are only approximations and fragments of the overall structure. Just as Maxwell's equations, on their own they may seem to be meaningless. They are like disconnected dots spread over an empty space. However, they are distributed according to a certain design. The design may be perceived when one learns how to quiet down the noise generated by intellectual reflexes and emotional reactions. Only then, one may become capable of recognizing and connecting the dots. By connecting the dots, one starts to see the emerging patterns. It is through experiencing these patterns that knowledge is imparted, because familiarity with the elements of the invisible worlds, however produced, enables the individual's mind to operate in a higher realm.

The Macrocosm

For where the beginning is the end will be.
(*The Gospel of Thomas*)

The words "I wished to be known" are the expression of the Absolute that led to creation. This expression defines the first stage of creation in which the Absolute presents itself as Pure Essence ("I"), Will ("wished"), and Pure Intellect ("to be known"). *Pure Essence* encompasses the divinity itself; *Will* is an unconditional identification with the purpose of knowing. *Pure Intellect* is invested with knowledge of the essence and of the process of knowing. Pure Essence, Will, and Pure Intellect are the original three pillars of the entire creation. In most scriptures, these three pillars are referred to as the original triplicity. They form the Realm, the highest cosmic layer. It is through this original triplicity that the Absolute initiated the process of creation.

These three pillars apply to the Absolute. They alone are capable of bringing into existence the invisible and visible worlds. However, they alone are not enough to produce the desired effect of "I wished to be known." In order for the Absolute's plan to be realized, there must be a recognition of the corresponding mode of triplicity on the part of the observer, who was to be located within the lowest strata of the creation. This means that the observer's inner structure would have to be also based on triplicity.

The purpose of creation can be realized only when the two triplicities, that of the Absolute and that of the observer, coincide. In other words, the thing in response to the creative impulse comes into being through its own act, not the act of the Absolute alone. This is the ultimate law of the creation. It is this law that, in its simplified materialistic form, was discovered by physicists as a new vision of the quantum universe where the observer, the observed, and the act of observation are interlocked.

The original triplicity of the creation percolates through the various layers of the invisible and visible worlds. It is transmitted and distributed via a field of universal consciousness. The field of consciousness is a single substance that, in varying degrees of refinement, permeates all creation. It is through this field that universal perfection, majesty, and beauty are manifested throughout all the worlds.

In accordance with the plan, the observer is located in the most remote regions of consciousness. Therefore, it was necessary to establish an infrastructure which would allow to accommodate such a huge gradient of consciousness. The overall structure is called the Cosmos. A higher (subtler) part of this infrastructure consists of invisible worlds, i.e., worlds which are not perceptible by physical senses. This part of the Cosmos is called the Macrocosm. The lower part of the Cosmos is referred to as the Universe, i.e., the physical world. This lower part is perceptible by the physical senses.

The Cosmos includes the worlds that have been formed as the result of several removals of the field of consciousness from the Realm. The following strata were formed:

- the world of ideas,
- the world of symbols,
- a buffer zone, and
- the world of phenomenal objects.

The *Realm* contains the original triplicity of creation.

The world of *ideas* contains multiplicity of the aspects which constitute the original triplicity.

The *symbols* are the ideas expressed as forms.

The phenomenal *objects* are the physical manifestations of the symbolic forms.

The *buffer zone* acts as a transition zone between the visible and the invisible.

As the field of consciousness cascades down from the Realm, it fashions the world of ideas. The world of ideas is the second strata of the Macrocosm, just below the Realm. It operates outside of existence. It is there that the set of primordial ideas and concepts appeared. The ideas and concepts may be thought of as centers within the finest layer of the field of universal consciousness. These centers constitute elements of a matrix, which contains all the possible attributes and aspects of essence, majesty, and beauty.

The various centers of the world of ideas form multiple sublayers. They are not "one;" they are "many." Each of these sublayers has its own name. These names describe the various attributes and aspects. They are just like reflections of an object in a multiplicity of mirrors. In one sense, all those names have an independent existence solely by the virtue of the generalized idea. The very fact of multiplicity provides the possibility of flaws. By being multiplied, they were exposed to a possibility of imperfection. This led to the appearance of some flawed states among the centers of absolute perfection. Despite these flaws, the world of ideas is to serve in its entirety as an active template for the lower worlds.

The cascading down of the field of consciousness led to the projection of the world of ideas onto a coarser level. The stratum below the world of ideas is called the world of symbols or the world of image-exemplars. Within the world of symbols, each of the centers of the world of ideas was projected as a multiplicity of forms of those various aspects and attributes. Like the world of ideas, the world of symbols consists of many sublayers. The symbols are needed, because the original ideas and concepts cannot be perceived within the limits of the physical world; they are too subtle to appear in the inferior world of phenomena.

As the field cascaded down, its degree of subtlety was discretely attenuated. As a result, the world of symbols is less subtle than the world of ideas. This cascading does not mean getting outside of the most subtle zone of consciousness. It is rather like a coarser layer being interwoven within a finer fabric. Both worlds are interwoven together and permeate each other. This feature applies to all the worlds, visible and invisible. They are all meshed together like a multilayered carpet. The difference between them is in their finesse and not in their location.

The world of symbols operates outside of the limitations of the ordinary time. Inherent in the world of symbols are forms of everything that is able to be, even before it actually comes into existence. This is just like the human capacity to visualize a desired action in the mind before that action is made manifest in the external world. It can be said that a form which lies hidden in the world of symbols is precisely the same as its reflection which is seen within the world of phenomena. The objects of the physical worlds are the likeness and reflections of the forms of the world of symbols.

The symbols represent all potentialities of all possible states, forms, relationships, and shapes of the entire physical world. The symbols are used to project the various aspects and attributes of the world of ideas onto the visible world. It is through the symbols that the qualities of the Absolute in their veiled forms can be manifested in the physical world. In the physical world, they are associated with various forms of physical beauty and perfection.

The various layers of the Macrocosm transcend the apprehension of the physical senses and the imagination. Although subtle, the strength of the field of consciousness is so powerful that it would annihilate any trace of matter. Therefore, there is no possibility for the existence of any matter there. Consequently, the Macrocosm is beyond the limitation of space, time, and existence. Yet, in accordance with the experiences of the mystics, the Macrocosm contains the original template for the entire physical world and the human mind. In other words, the Macrocosm holds the design for the entire creation, from its beginning to its end:

> When you see your likeness, you are happy. But when you see your images that came into being before you and that neither die nor become visible, how much you will bear!
> (*The Gospel of Thomas*, *84*)

Of course, the mystical model of the cosmic structure is only an approximation that is expressed by means of allegories and symbols. Even though the mystic's experience cannot be verified by the criteria of materialistic science, it is worthwhile to reach towards it for it provides indications about the overall structure of the physical world.

The world of symbols is separated from the phenomenal world by a buffer zone. The buffer zone is like a veil that surrounds the entire Universe. It makes the Macrocosm imperceptible to the physical senses. Science has recognized the presence of such a buffer and identified its function as the veiling of reality. Just as indicated in the first Chapter, "outside time and space is inconceivable, and unfiltered reality would blow the brain's circuits."

The Realm, the world of ideas and the world of symbols form the Macrocosm, i.e., the intelligible worlds that are beyond physical existence. Sometimes, the various strata of the Macrocosm are referred to as "heavens."

The next stage of the process of creation is realized when the world of symbols is projected through the buffer zone onto the next lower layer. Again, the subtlety of the field of consciousness is further discretely attenuated as it is cascading onto the crudest form of its manifestation: the physical world. This attenuation is qualitative in its character. It is needed to accommodate the gradual descent of consciousness onto the lowest world.

The physical world (the Universe) occupies the lowest position within the cosmological structure. The Universe and each of its constituents are all intertwined within the Macrocosm; they are permeated by the various layers of the field of universal consciousness. This means that the entire Universe, including every form of matter and life, constitutes the coarsest layer of the field of consciousness that is interwoven within finer layers of subtler consciousness. It is in this sense that the Cosmos is a gradient of consciousness and on this gradient the physical world occupies a low level.

It is through the various levels of the field of universal consciousness that the physical world is linked to its original source. This is alluded to in the Gospel of John by the statement:

> In the beginning was the Word
> (*John, 1:1*)

"The Word" is a symbol. This indicates that this particular symbol (the Word) appeared within the world of symbols. If the "Word" was formulated within the world of symbols, it would have had to have its origin within the world of ideas. Are there any indications what that idea was?

It turns out, that there is such an indication. It is expressed in the Koranic verse that is referred to as "the heart of the Koran." The verse is entitled *Ya Sin*. *Ya* and *Sin* are two letters of the Arabic alphabet (Y, S). These two letters indicate the link between "The Word" and its originator within the world of ideas. The link was explained by Hakim Sanai, a 12th century Persian poet, in the following verse:

> From *Kaf* and *Nun*, like a precious pearl,
> He made the eye into a mouth filled with *Ya Sin*.[9]

The mystics have their own language which they use to communicate their experiences. Their language is based on the *abjad* system. In the abjad system, letters of the alphabet are given numerical meaning. Therefore, mystical phrases can be translated into numbers and vice-versa. In this manner it is possible to describe the links between ideas (phrases), symbols (words), and things. The mystics use this system to convey their experiences of the relationships between the physical world and the invisible worlds.

This system is often used in Arabic and Persian poetry. Some of the fables and stories are written in the abjad code. Although the abjad is inherent in the design of the Arabic language, some

---

[9] *The Walled Garden of Truth,* Hakim Sanai; translated by David Pendlebury (The Octagon Press, London, 1974, p. 56).

simplified examples of its application may be translated into other languages. A few examples of its application are found in English poetry.[10]

According to the abjad system, the letters *Ya* and *Sin* are equivalent to the letters *Kaf* and *Nun* (K, N). The letters *Kaf* and *Nun* form God's primordial command: "Be!"[11] This command is spelled out in the Koranic verse as:

> Be, and it is!
> (*Koran, 36:82*)

In the context of Sanai's couplet, the "eye" indicates the precursor and the originator of the "word." The "eye," the consciousness at the level of the world of ideas, recognizes "a precious pearl" ("the hidden treasure"). At this level, the hidden treasure is formless. Afterwards, this idea is passed over onto the world of symbols, where it takes on the form of "the Word." This word is: "Be." From the world of symbols, it is manifested as the physical world ("Be, and it is!")

This is one of the examples of how the Bible and the Koran are complementing each other. This relation, however, remains hidden if one limits oneself to the literal interpretation of the scriptures.

The command "Be!" triggered the event, which the modern physicists coined the Big Bang. The Big Bang was the coming of the physical world out of the invisible; like the light coming out of

---

[10] Shakespeare was familiar with this system. He used the abjad system in the episode with Scarus in *Antony and Cleopatra* and in the Dedication to the *Sonnets*.
[11] The letters *K* and *N* form the root of the word *kun* (be).

the darkness. This is alluded to in *Genesis* as the appearance of the Day out of the Night:

> And God said, Let there be light: and there was light.
> And God saw the light, that it was good: and God divided the light from the darkness.
> And God called the light Day, and the darkness he called Night. And the evening and the morning were the first day.
> (*Genesis 1: 3-5*)

This was the first stage ("the first day") of the process leading to the appearance of the Universe. It was the beginning of everything that came into physical existence.

The intention "to be known" led to the creation of the physical world. It was this intention that was the driving force behind the entire process. The physical world was to provide an environment within which the jewel ("precious pearl") could be hidden. It was to be hidden by being veiled by the multiple layers of the worlds of ideas and symbols. In the overall scheme of things, man was to be the host of the "jewel" that was hidden within himself. Again, it should be kept in mind that these various layers are not localities; they are the various states of the human mind.

The various layers of the Macrocosm correspond to various stages of higher consciousness. In order to reach towards the Macrocosm, man must overcome the limitations of space and time. Therefore, man was granted some means which allow him to break through these limitations and discharge correctly his evolutionary obligation. However, these means were invested in man in their latent forms. It may be said that these "devices" have been veiled. The veils form a barrier, i.e., a buffer zone. It is the existence of this buffer zone that has been identified by scientists as "the principle of veiled non-locality." The buffer zone divides the invisible worlds and the physical world. In a natural or an ordinary state, man is incapable to overcome this barrier. The barrier may be overcome only through directed and conscious efforts. The activation and development of these latent devices is a crucial step of the entire process. It is in this sense that man is the most vulnerable link in the entire cosmological structure. It is this challenge that determines the fate of humanity.

The Universe was created as a shell within which mankind was to come into existence. By familiarizing ourselves with the various cosmic strata, we may get a hint about the structure of the human mind. In other words, the various stages of the development of the inner layers of the human mind are reflections of the Macrocosm.

The Dot

> Geometry will draw the soul toward truth and create the spirit of philosophy.
> (*Plato*)

It is late summer 2013 in Trieste, an Italian city on the shores of the Adriatic Sea. One of the sessions of the International Conference on Supersymmetry and Unification of Fundamental Interactions is taking place. The conference is devoted to new ideas in high energy physics.

We are inside a magnificent conference room.[12] Nima Arkani-Hamed, an American-Canadian theoretical physicist, is on the stage in front of the audience. He is wearing a dark blue shirt loosely hanging outside his short-pants. He walks with a steady pace from one side of the stage to the other, and then he turns and walks back. Like a pendulum swinging across the stage, he tirelessly repeats his movements. Behind him, there is a large screen. On the screen there are odd looking images. They are like doodles, filled with arrows, polygons, numbers, circles, and other not clearly defined shapes and curves. Each time he passes in the front of the screen, a new piece is added to the previous image and a new doodle appears. For an outside observer, this looks like a man who is building a new incredibly

[12] *The Amplituhedron* (https://www.youtube.com/watch?v=q4Dj8fq30sk).

sophisticated structure. Just like the sophisticated bower of Architect, the bird-of-paradise; with each return, he brings a new twig to his "bower." But then, this structure is placed in multidimensional space and takes the form of a very elaborate geometrical figure. Encoded in its volume are the most basic features of reality that can be arrived at through mathematics. They are called "scattering amplitudes." They represent the likelihood that, upon colliding, a certain set of particles will turn into other specific particles. Instead of traditional mathematical equations, this crystal-like multidimensional geometrical figure represents a sophisticated operator. This operator is one of the latest intellectually formed structures that may represent the "soul" of matter.

***

Scientists believe that the origin of everything may be sufficiently described by a set of complex mathematical relationships that can be reduced to a single equation. Such an equation would be expressed in the following form:

$$\{\}\Psi = 0$$

The symbol $\{\}$ is an operator that encompasses all the mathematical functions Ψ needed to describe all the relationships between the unified forces and the complete set of the most elementary particles. Everything that exists will be included and determined by the operator $\{\}$.

Mathematics acts as an operator that allows expressing "things" as symbols. The "things" are quantifiable; therefore, they are always equal to a number. In this way it is possible to describe the quantitative relationships between "things" whose properties are represented by numbers. This is why, on one side of the equation there is a "0." The "0" indicates that all effects and things have been included, accounted for, and balanced out.

The operator {} remains unknown. It has not yet been discovered. It is possible, that instead of a set of lengthy mathematical formulas, the operator will take the form of a sophisticated multidimensional geometrical "bower." Finding and formulating it constitutes the ultimate goal of science. It is thought that when it is found, then everything will be known about the Universe and life. The operator is the central point behind the idea of "the theory of everything," i.e., a hypothetical theoretical framework of functions that intends to fully explain and link together all physical aspects of the Universe. The "theory of everything" is like the ultimate prediction machine - a single equation from which everything follows. Although scientists exclude "divinity" from their inquiry, they often like to make references to a divine factor. In this particular case, they believe that the operator {} would be a substitute for "the mind of God."

***

While the scientists have concentrated their efforts on understanding how the physical world came into existence, the mystics were looking into events that took place before the appearance of the physical world. They looked towards the Macrocosm. They worked with an equivalent formula to the above "bower" equation. In the symbolic language of the mystics, the formula is expressed in the following form:

I am that I am
(*Exodus: 3-14*)

The left part of the formula, "I am that," encompasses the entire physical world. This part is an equivalent of the "bower" equation. The remaining part of the expression, "I am," indicates the effect of the Absolute's presence. According to the mystics, "I am" is the factor that keeps the Cosmos in balance. Without it, the formula would be incomplete. The entire Cosmos is the simultaneous manifestation of these two states, "I am that" and "I am." At every moment, these two states determine the complete state of the Cosmos.

This formula becomes more obvious in its Koranic version:

No God but God
(*Koran, 47:19*)

In the Koran, the physical world is referred to as "No God." As in the Biblical expression, the physical world is kept in balance by "God."

Now it is easy to comprehend the position in which science has placed itself. Science is concerned only with the left side of the mystical formula. In this context, the scientific approach may be expressed by the following equivalence:

(I am that) $\equiv 0$

This equivalence excludes the presence of the Absolute. It is like a bird with a single wing. Such a bird cannot fly. Although balanced within the phenomenal world, the formula is incomplete in the context of the overall Cosmos; it misses a key factor.

Similarly to science, religions are focussed only on one side of these formulas. They are concerned with the right side, i.e., "God." In religious approach, man, as an active participant in the creative process, is disregarded.

The mystics, on other hand, took into consideration the entire sequence of the formula. They have realized that the secret of the creation and the purpose of life cannot be expressed adequately by numbers. It can, however, be indicated by allegories and symbols.

The mystics have noticed that there is something intentionally missing in the Biblical and Koranic formulas. This "something" is needed to enforce the balance, symmetry and beauty of these expressions. Therefore, they focussed their attention on this missing element. Then they were able to perceive that there is a hidden "dot" in the middle of this seemingly paradoxical expression:

No God • but God

It is this hidden "dot" that fuses together the visible and the invisible. Just like the point in the middle of the symbol of infinity:

$$\infty$$

The left side of the mystical expression applies to the physical world. The right side represents the Macrocosm. The "dot" is like a hidden door that leads from the world of phenomena to the Macrocosm.

The "dot" is a symbolic indication of how the secular and the sacred, the scientific and the mystical, the sensual and the visionary, the imaginary and the real - are always mingling together. In this context, the "dot" may be thought of as an operator. This invisible operator provides a framework allowing man to develop his mind so he becomes capable of experiencing the working of the invisible worlds. It is only through experiencing the invisible worlds that it is possible to find satisfactory answers to questions about the purpose of the Universe and human life.

The abjad system may further help with decoding the meaning of the "dot." The Arabic root for the word "dot" contains three letters Q, N, and T. These letters are equivalent to the number 159 (100, 50, 9). The number 159 may be expressed as (40 + 5 + 50 + 4 + 60). These numbers are equivalent to a set of consonant letters (M, H, N, D, and S). These letters form the root of the word *muhandis* which means "Architect."[13] So, the word "dot" is a symbol indicating "Architect." If one extends this meaning to the concept of the Prime Architect, an aspect of the Absolute, then the "dot" acquires a dynamic function which allows one to move from the physical world towards the invisible worlds of symbols and ideas. It indicates the starting point of the journey towards the Absolute. It is there that the *journey* leading to "beyond" starts.

---

[13] *The Sufis*, Idries Shah (The Octagon Press, London, 1964, p. 372).

The abjad system is thought of as a precursor of mathematics. The difference is that mathematics cannot perform operations on infinite numbers. All mathematical symbols belong to the world of phenomena. Therefore, mathematics cannot be applied to the relationships within the Macrocosm; mathematics is limited to the physical world. With the abjad system, however, it is feasible to do so. For example, the "dot" in the middle of the paradox is not only marking the way to the invisible. It also conveys a part of the developmental methodology. Namely, when taken separately, the two letters (Q, N) of the word "dot" mean "deep meditation." In order to reach "beyond," one has to focus one's attention on the "dot." It is there that the inner secret may be experienced. This is indicated by the remaining third letter (T), which stands for "inner knowledge." In other words, the "dot" is a pointer indicating not only the ultimate destination of the journey. It contains instructions on how to start the journey. The journey starts with a "meditation" that leads to "inner knowledge." Inner knowledge is gained through the development of inner layers of the mind. By developing these layers, which are known as inner faculties, one may arrive in the presence of the Prime Architect.

For the mystics, the "dot" marks the beginning of their journey and also its final destination. It indicates the purpose of life. As a matter of fact, it marks the beginning of a new life. Its function is illustrated in many stories and fables as a magical cave with treasures, a magical horse that rises to heavens, a genie in a bottle, a ring that opens the door to a magical kingdom, etc., etc. In this context, the "dot" is the mystical equivalent of the universal operator that is being hunted by the physicists. But there is a great difference between these two "operators." The scientific operator is to be formulated in the language of mathematics. The "dot," in order to be functional, has to be experienced. Of course, one would have to be familiar with the overall concept and ideas that underline the "secret" that is hidden in this seemingly trivial symbolic representation. Just as in the case of Maxwell's equations, for a person who is not familiar with this specific "science," this symbol neither has meaning nor importance.

Somehow, this "dot" has been imprinted in nature in creatures such as the bird-of-paradise and his magnificent bower. In this context, the bowerbird has been overdesigned. He does not need to beautify his bower; he just can't help doing it. His activity, however, may help man to start to ask himself: why does the bird do it? And that's the only reason for having the bird and his bower. The bird-of-paradise's fixation with beauty is a hint about the cosmic truth that is inherent in the physical world. Physical beauty is but an indication of cosmic perfection. It is an ever-present reminder that, just like the bird's brain, the human mind in its natural ordinary state is attracted to it but is incapable of fully perceiving its meaning. By limiting one's attention to the physical manifestation of beauty and focusing one's efforts on its reproduction, one diverts oneself from the primordial intention that is the prime driver of human life. Sooner or later, this type of beauty vanishes; it is perishable. However, unlike the bird's brain, the human mind can overcome its limitations. It can be prepared to overcome the illusionary perspective of reality and perceive the meaning that is encoded in natural beauty.

Science, just like the bird-of-paradise, is driven by an intuitively detected beauty and the perfection of forces ruling over matter. It feels an incredible attraction towards it, and it is compelled to reproduce it by applying the apparatus that it has developed, i.e., mathematics. In this sense, the scientific approach is a copy of the bird-of-paradise and his fixation with the "bower" - as it tries to "beguile nature of her custom."[14] As a result, science ends up like that bird, being locked up in endless building and decorating - a spectacular but developmentally sterile activity. This situation is illustrated in the tale "Zaky and the Dove":

---

[14] See Note #7 on page 18.

THERE was once a man named Zaky. Because of his capacities and his promise, a certain teacher -the Khaja- decided to help him. This Khaja assigned a subtle creature of special powers to attend upon Zaky and to help him whenever he could.

As the years passed, Zaky found that his affairs prospered. He did not imagine that such advantages as he was receiving were entirely due to himself, and he started to notice a coincidence of events.

Whenever his affairs were about to go well, he observed, a small white dove was to be seen somewhere nearby.

The fact was that the subtle attendant, in spite of his powers, needed to be within a certain distance of Zaky to carry on his work. In spite of his remarkable abilities, in his transition into the present dimension he had to take a form. A dove was the form which he had adopted as most suitable.

But Zaky only connected doves with luck, and luck with doves.

So he started to keep doves, and put down food for any dove which he saw, and to have doves embroidered on his clothes.

He became so interested in doves that everyone in the world thought of him an authority on them. But his material and other affairs ceased to prosper, because his concentration had been diverted from intention to manifestation, and the subtle attendant in the form of a dove had to withdraw, to avoid becoming the cause of Zaky's undermining himself.[15]

The physical world is moulded in accordance with the design of the Macrocosm. The field of consciousness is the driving force of the

---

[15] "Zaky and the Dove" included in *The Magic Monastery*, Idries Shah (The Octagon Press, London, 1981, p. 72).

entire process. In the physical world, the driving force is manifested in a multiplicity of forms. Over the centuries, these various forms have been gradually discovered as laws of physics and rules of mathematics. These laws and rules allowed to formulate the various theories and models of the Universe. In this context, one may look at the laws of physics and mathematics as Mannerist projections of the principles residing within the world of symbols. The world of symbols contains the entire template of the physical world; it is this template that science attempts to decode. In other words, such an attempt corresponds to an effort towards reaching the world of symbols via ordinary intellect, an impossible task. The laws of physics formulated by science are in reality skewed and fragmented elements of the matrix which operates within the world of symbols. The skewedness of the laws of physics is like the "maggot in the apple" in the tale "The Celestial Apple":

> Ibn-Nasir was ill and, although apples were out of season, he craved one.
>
> Hallaj suddenly produced an apple.
>
> Someone said: 'This apple has a maggot in it. How could a fruit of celestial origin be so infested?'
>
> Hallaj explained:
>
> 'It is just because it is of celestial origin that this fruit has become affected. It was originally not so, but when it entered this abode of imperfection it naturally partook of the disease which is characteristic here.'[16]

---

[16] "The Celestial Apple" included in *The Way of the Sufi*, Idries Shah (The Octagon Press, London, 1980, p. 256).

This is why familiarity with the Macrocosm provides important clues which may guide science and scientists in their attempts to describe the modus operandi of the Universe. The interesting thing is that some of the most recent discoveries of quantum theory are partial but accurate reflections of the mystical experiences. As such they allow to express some of the mystical experiences in the language of modern science. Even if the mode of acquiring mystical "data" is not acceptable to science, the mystical experiences may serve as a useful reference for the most recent scientific discoveries. Furthermore, mystical "data" may help to make science more effective by indicating directions towards more adequate approaches and targets.

It is for this reason that rational science should reach towards the mystical data base as a source of relevant information needed for an effective and meaningful continuation of the exploration of life and the Universe.

The Cosmic Oscillator

> If you want to find the secrets of the Universe,
> think in terms of energy, frequency and vibration.
> (*Nikola Tesla*)

As the field of universal consciousness was cascading down from the world of symbols, it formed a barrier. This was the buffer zone between the world of symbols and the physical world. At the buffer zone there is a major degradation of the field of consciousness. The Universe was to be formed within the grossest zones of consciousness.

The concept of the physical world was first conceived within the world of ideas. Afterwards, the template of the entire Universe appeared within the world of symbols. The template contained all the possibilities of everything that has ever existed, could exist or will exist in the Universe. In the world of symbols, the template appeared first as the "dot," i.e., a *non-dimensional* point. When projected from the world of symbols onto the crudest mode of consciousness, the "dot" acquired one dimension. The Universe was to be born from that *one-dimensional* point. The projection of the "dot" into the one-dimensional point corresponds to the transformation of the lowest grades of consciousness into matter: it marked the birth of matter. This event has become known as the Big Bang.

According to modern science, the Universe was created as the result of the Big Bang. Although the current model of the Big Bang is still at a hypothetical stage, so far it offers the best available approximation that helps to grasp the overall concept of the creation of the Universe.

Prior to the Big Bang, the entire Universe was contained within the one-dimensional point. The one-dimensional point is the lowest level of the entire cosmic creation. Its appearance marked the completion of the descending part of the loop of creation.

The boundaries of the future Universe were to be manifested in the phenomenal world by the appearance of a medium which science coined as space-time. The entire space-time was to be enclosed between this one-dimensional point and the buffer zone. In other words, the region of the field of consciousness between the one-dimensional point and the buffer zone determined the boundaries of the future Universe. The conditions were ready for the initiation of the ascending part of the loop. The Big Bang marked the commencement of the ascent; the Big Bang was the first step of the ascent towards the Absolute. No wonder that humans are so fascinated with this event.

The first phase of the ascent may be called the mechanical creation. Although divinely conceived and implemented, it was strictly a "mechanical" process. Space, time, and matter were the products of the first phase of the mechanical creation. The process was limited by a number of restrictions that were imposed on space, time, and matter. These limitations were dictated by the overall purpose of the Universe, which was to provide sufficient conditions for the fulfilment of the evolutionary potential that was invested in mankind. Later on, the space, time, and matter restrictions were gradually discovered as laws of physics.

The size of space was so determined as to ensure sufficient conditions for the appearance and support of life; matter determined the amount of raw energy needed to create and sustain life; time was determined in such a way as it would be sufficient for the fulfilment of man's function.

Matter was formed within the lowest degrees of consciousness. It was the lowest grade of consciousness that would lead to the formation of minerals, vegetation and animals.

The mass needed for the formation of the entire Universe was invested in the one-dimensional point. As a result, density and the curvature of this one-dimensional point would have been infinite. Such a point is an example of what mathematicians call a singularity. Any mathematical theory breaks down at such a point. This means that mathematics cannot be applied to any events before the Big Bang.[17] Science attempts to understand only what has happened since the Big Bang. As far as science is concerned, events prior to the Big Bang have no consequences on their models.

As described by Einstein's famous equation ($E = mc^2$), mass is equivalent to energy. Therefore, the total energy of the entire Universe was confined within that singular point. As a result, the point became infinitely hot. It was then, that this singular point shattered into the infinitesimal precursors of the elementary particles of matter. This marked the birth of the Universe. This was also the beginning of space, time, and mass. The space-time-matter defined the limits of the material world.

While attempting to build a model of the Universe, scientists concluded that all particles are in fact waves. Consequently, everything within the physical world, including the Universe itself, has its own probability wavefunction. This means that we are living in a world of quantum possibilities. According to quantum theory,

---

[17] It is interesting to realize that it is not a philosophical doctrine but simply the limitation of mathematics that imposes on science a belief in a completely self-contained Universe.

the probability of an event only exists if there is "someone" to observe it and experience it. Only then, the wavefunction may collapse and manifest itself as a particle. This is one of the basic laws of quantum physics. It seems somehow obvious that the divine principle of the "observer" is encoded within the laws governing the physical matter. What is extraordinary, however, is the fact that this law was discovered by modern science and became one of the fundamental elements of quantum physics.

Quantum theory implies that in order for the Universe to be manifested, there has to be an observer that can observe it so as to collapse its probability wavefunction. Of course, in the early stages of the Universe there was "no one" within the Universe to observe what was happening. Therefore, the wavefunctions of the initial particles could not collapse; the Universe could not come into being.[18] This clearly indicates that such an "observer" could exist only within the higher zones of consciousness, i.e., outside the space-time-matter limitations. This is in compliance with the rule (X + 1) which states that the act of "observation" of a level X may effectively be executed by an observer acting within the level of consciousness (X + 1) or higher. This rule reflects the fact that the Cosmos is a gradient of consciousness. Consequently, an ordinary man by the act of observation may collapse the wavefunctions of photons and elementary particles. An ordinary man, however, cannot collapse his own wavefunction. For man to appear, an "observer" within the Macrocosm is needed.

The Universe was created so man could fulfil his evolutionary function. The entire system is designed in such a way that man's actions are reflected within the various cosmic strata. This implies that the overall system is dynamic; it is changing in accordance with man's actions. Following man's actions, resulting changes are being made within the Macrocosm. In turn, these changes are projected

---

[18] It is at this point that the *loop of self-consistency* proposed by science (referred to in the first Chapter) breaks down.

back onto the physical world. In other words, there is an active feedback between the physical world and the invisible worlds. This means that the Universe is being continuously updated. Such a dynamic state of the Universe is referred to in the scriptures. Psalm 33 alludes to this fact:

> By the word of the Lord were the heavens made;
> And all the host of them by the breath of his mouth.
> (*Psalm, 33:6*)

This *Psalm* implies that the Universe with its galaxies, stars and planets - was made as a result of the Lord's breath.

Jami, a 15th century Persian poet, further elaborates on this concept. In his *Flashes of Light,* he states that the Universe is sustained by "breathing":

> This universe consists of accidents all pertaining to a single substance, which is the Reality underlying all existences. This universe is changed unceasingly at every moment and at every breath. Every instant one universe is annihilated and another resembling it takes its place.[19]

Every instant the Universe is annihilated and another resembling it takes its place. With each "breath," it is born and dies; then it is born again. Expansion and contraction are taking place at the same

---

[19] *Flashes of Light*, Nurudin Jami; translated by E.H. Whinfield and Mirza Kazvini (Royal Asiatic Society, London, 1906, p. 42).

moment. This suggests that the Universe, as a part of the Cosmos, is not permanent. It constitutes a sequence of "accidents;" ever changing and being renewed at every "breath." At each instant, the Universe is disappearing and is being replaced by a new one. As a result of these rapid successions, the spectator is deceived into the belief that the Universe has a permanent existence:

> Every moment the world is renewed, and we are unaware of its being renewed whilst it remains the same in appearance.
> (*Mathnawi, Book I, 1144*)

With each "breath," the Universe is completely erased and remoulded into a new one. A new Universe is born. This erasing and remoulding is taking place simultaneously. It is in this way that changes within the Universe and the Macrocosm are instantaneously transmitted to each other. In other words, the "breathing" allows for the establishment and maintenance of a link between the physical world and the invisible worlds; with each new "breath" the link is renewed. We may now realize that, through "breathing," the Universe is permanently entangled with the various strata of the Macrocosm. This is why it is necessary for the Universe to be erased and simultaneously moulded into a new one.

It is quite interesting to note that such a feature of the Universe's behaviour has been confirmed by modern physics. It is related to the discovery of the effect of quantum entanglement which, quite appropriately, has been coined by Einstein as "spooky action at a distance." Quantum entanglement allows for a number of particles to behave as one, regardless of how apart they are. Actions performed on one of them seem to be instantaneously influencing the others. This is taking place without any physical communication between them. In other words, through entanglement it is possible to transfer instantaneously key properties from one system to another. However, this effect may take place only when these systems (particles) are correlated in a very specific way. Quantum devices, called entanglers, produce entangled particles by "breathing" them out at the same time. In this way they gain inner coherence; they become entangled.[20] This was probably the first example where the combination of a mystical concept and theoretical research led to the birth of a new field of modern technology: quantum photonics. Quantum photonics provides the basis for the development of quantum communications, teleporting devices, quantum cryptography, and quantum computers.

In this context, the entire Universe may be compared to a giant oscillator. It oscillates between its initial state of being a one-dimensional point and the state of being a wave field containing the entire Universe with its galaxies, stars and planets. At any one moment, the Universe manifests only that part of itself that is within available possibilities. In other words, we are stepping out from an old universe and stepping into a new one, which is statistically the most probable one.

Each new Universe is slightly different from the previous one. It is different because it grows. It grows in accordance with the currently operating template projected from the world of symbols. Its growth is marked by two features, the range of the oscillations and the composition of the newly appearing physical structures (such as supernovae, black holes, white dwarfs, red giants, etc., etc.). The range of the oscillations defines the size of the Universe. The variety of appearing structures marks its maturity. The Universe will be increasing till it reaches its maturity.

---

[20] *Applied Microphotonics*, W. Jamroz, *et al.*, (CRC - Taylor & Francis, Boca Raton, FL, 2006, p. 284).

In the framework of the overall cosmological structure, appearances and disappearances of universes are a reflection of changes within the world of symbols. There is nothing in the phenomenal world that does not have its original form in the world of symbols. In other words, the entire time sequence has to comply with the design constructed within the Macrocosm. This means that before the universe came into existence, its fate was precisely determined.

The Universe is like a corporeal body placed temporarily within the incorporeal Macrocosm. Within the Universe, time has its present, past and future. From the perspective of the Macrocosm, these three states are in reality one. The world of symbols is outside of the limitations of ordinary time. A symbol contains the entire developmental sequence of a form which appears in the physical world. As a symbol is being projected into the physical world, its multiple physical manifestations unfold themselves partially in space-time. This partial unfolding is displayed as series of physical births and deaths of this particular symbol's sequential versions:

> Every form you see has its archetype in the placeless world;
> If the form perished, no matter, since its original is
> everlasting.[21]

During their physical existence, these unfolding versions are being adjusted according to their original template in the world of symbols. At every moment every single physical form has to be remoulded to fit gradually into its corresponding original symbol.

---

[21] *Divan-e Shams-e Tabrizi*, Jalaluddin Rumi; translated by R.A. Nicholson (Ibex Publishers, Inc., Bethesda, MD, 2001, p. 47).

***

The field of universal consciousness creates the material Universe; it defines the boundaries of the physical space-time. The field is not static; it oscillates and expands in accordance with the Cosmos' "breathing." Because of the space-time boundaries, the oscillations in the form of waves are reflected back and forth. As a result, they form a set of standing waves. Within fixed boundaries, a standing wave is a wave whose amplitude is stable and does not move. The locations at which the amplitude of the standing wave has its minimum are called nodes, and the locations where the amplitude has its maximum are called antinodes. The standing waves reflecting back from the edge of space-time form various patterns. As a result, the entire field of consciousness is divided into regions bounded along nodal lines. The sizes, shapes and locations of the regions are determined by the boundaries of space-time and the "frequency" of the oscillations. When the frequency of the oscillations increases, more sophisticated patterns occur. These patterns are moulds for various forms of matter.

The entire process is similar to the formation of Chladni patterns.[22] In quantum mechanics, Chladni figures ("nodal patterns") have helped to find the solutions of the Schrödinger equation, i.e., a differential equation that describes the wavefunction of a quantum-mechanical system. The mathematics describing Chladni figures was used to arrive at the understanding of electron orbitals. It was this process that led to the realization that matter may appear either as a particle or a wave.

In the case of the field of universal consciousness, each of the nodal regions contains a template of complementary forms and properties of "things." The appearance of matter is the result of the collapse of these various nodal patterns. The presence of physical matter is the confirmation of the existence of space-time

---

[22] Named after Ernst Chladni (1756 – 1827), a German physicist and musician.

boundaries. The boundaries are needed to form the nodal patterns; without the boundaries matter could not be formed.

According to the standard theories of cosmology, the Big Bang took place about 13.8 billion years ago. At first, the physical energy was captured by the first set of nodal regions within the lowest zone of the field of consciousness. Upon capturing the energy, the simplest nodal regions collapsed and formed the elementary particles, the first physical layer of matter. The shapes and properties of these particles were determined by corresponding templates within the world of symbols. In other words, the matrix within the world of symbols acted as the "observer" of the Universe. The very early Universe was formed. It is estimated that this very early Universe was formed within the first picosecond (one trillionth of a second) after the Big Bang. This period is referred to as the Planck epoch. The physicists believe that, at that time, the four fundamental physical forces that shape our Universe, i.e., gravity, electro-magnetism, weak nuclear forces and strong nuclear forces were combined and constituted one unified force.[23] One second after the Big Bang, the Universe contained mostly photons, electrons, protons and neutrons. These particles became the building blocks of atoms and molecules.

Afterwards, a second series of oscillations within the lowest zone of the field of consciousness was initiated. These oscillations were superimposed on the first ones. The overlap led to more sophisticated nodal regions. This time, the nodal regions of consciousness formed the templates of the smallest atoms. It is estimated that this second switching happened about one hundred seconds after the Big Bang. It was then that protons and neutrons

---

[23] Devising a model of such a single unified force is perhaps the greatest goal of today's theoretical physicists. In reality, however, the fundamental forces were united within their original source that was placed in the world of symbols, i.e., outside of the physical world. They appeared as multiple forces when they were projected into space-time. This is why it will be difficult for physicists to construct a precise model of a single fundamental force.

started to combine together to produce the nuclei of the smallest atoms. Hydrogen and deuterium came into existence. These were the first and simplest elements. They were the precursors of the entire set of chemical elements.

The next increase in frequency of the oscillations of the field of consciousness took place about 3 minutes later. At that time, the protons and the neutrons were fused into heavier elements, mainly isotopes of helium. Within a few hours after the Big Bang, the production of helium and other elements stopped. For the next four hundred thousand years or so, the Universe kept expanding until it became cool enough for atoms to release photons. These early photons can still be detected today as cosmic microwave background. This microwave background is the oldest remnant of the Big Bang ever observed.

Afterwards, as the frequencies of the oscillating field of consciousness discretely increased again, a new set of nodal regions with more advanced structures was formed. This time, these new "compartments" carried the forms of heavier atoms. Their formation and appearances followed a specific pattern that was discovered by some scientists and then reproduced in the form of the periodic table. Then again, as the frequencies of the oscillations of the field further increased, the newly formed nodal regions provided the "traps" for molecules and more elaborate compounds. In other words, the oscillating field of universal consciousness provided the templates for the entire physical world, with its galaxies, stars, and planets. The Sun was formed some five billion years ago out of a cloud of rotating gas. The heavier elements of the cloud were trapped in the nodal zones which provided the templates for the planets that orbit the Sun. The Earth was one of these planets.

The currently observable Universe is contained within a fraction of the lowest zone of the field of universal consciousness. The region of its oscillations extends from the initial point to its current size. The limit of the size of the Universe is not detectable yet today; the Universe is still expanding.

The entire zone of the lowest field of consciousness with its nodal regions may be compared to a jigsaw puzzle that is in the shape of a multidimensional box. Each nodal region within the jigsaw box is like an empty compartment. At every new "breath," new pieces come into existence by collapsing within the available compartments. This means that with each breath only a fraction of the box is being populated. It is this part of the Universe that is observable. At the next breath, as space-time expands, more particles and compounds are available and more compartments are being filled in. This process will continue till all the compartments are fully occupied. At that future time, the Universe will reach its maturity.

The jigsaw box is the physical projection of Pure Essence. The surrounding field of consciousness may be thought of as the "soul" of the Universe. All created things are the outcome of an interchange between the two principles of essence and consciousness. They are strongly attracted to each other; the "soul" recognizes the beauty of "essence." When they are united, then a new form comes into being. One may look at the jigsaw box, the surrounding field of consciousness and the created things as a physical form of the original triplicity.

It would be difficult to imagine the shape of such a box, particularly its internal compartments. Their shapes are the result of quite sophisticated interactions between a series of standing waves and the spatial and temporal boundaries of the box. However, new developments by theoretical physicists may help us to have a better grasp of the shape of such a box. Namely, as was described earlier, the amplitudes of certain particle collisions may be encoded within the volume of a crystal-like geometric object. Encoded in its volume are the most basic features of the physical world that can be calculated, which represent the likelihood that a certain set of particles will turn into certain other particles upon colliding. It is in this sort of shape that the compartments are formed within the lowest zone of the universal field of consciousness. These particular compartments would be attracting the lowest layers of matter. More complex layers, such as atoms and molecules would be attracted to more sophisticated shapes. We can imagine that the entire Universe may be represented by a huge crystal-like living structure with an incredibly sophisticated multidimensional inner design.

***

The critical stage of the creation was initiated at the moment of the appearance of mankind. This was the end of the mechanical (involuntary) creation. The appearance of mankind was governed by the same laws as in the case of the Universe; the laws complying with the statistics of quantum possibilities. Mankind corresponds to the highest nodal zone in the Universe. The probability for the formation of such a zone is extremely low; so low that it allows for only one occurrence. This zone is confined to the planet Earth. This particular frequency of the field of consciousness does not appear anywhere else in the Universe. It is unique and specific to the Earth.

The purpose of the creation of the Universe was to provide the needed conditions for the appearance of mankind. If the jigsaw box was a kind of universal map, the appearance of man would be indicated by a specific location on that map. The fact is that man appeared on the map before the box was filled in, i.e., before the entire map was drawn. Yet, the laws of statistical physics require that the entire box be filled in completely before the Universe enters into its declining phase. This is why the expansion of the Universe did not stop with the appearance of man. (This may be compared to the purchase of the winning ticket in a lottery; the fact that the winning ticket has been bought does not terminate the selling of the tickets; it is the sale of the remaining tickets that will ultimately determine the value of the winning ticket.) This additional period provides some safety measure, a margin for errors. Taking into account man's natural characteristics, it was highly probable that he would not comply completely with his evolutionary obligation; he would not be able to fulfil his potential within the minimum prescribed time. After all, the process would not make sense if a certain degree of free-will was not warranted. In this context, the available quantum possibilities are a measure of man's free-will.

The overall cosmic design has to accommodate both, causality and free-will. Causality, i.e., cause and effect at the level of ordinary man, is the field of operation for the Will of the Realm. Any adjustment to the evolutionary plan is manifested on the level of ordinary man as a series of intertwined opportunities that appear in different places and at different times. But it is up to man to choose. There is no warranty that man will discharge his function correctly within the allotted time.

At one point, however, regardless of man's evolutionary state, all the possibilities within the jigsaw box will be filled in. At this point, the Universe will start to disappear. Starting with the most complex structures, everything will start to disappear to nothingness. All matter will disintegrate. The Universe was born, developed, and will gradually diminish.

The question may be asked: will mankind manage to discharge its function within its allotted time? Is it possible for us to gauge the status of the process with respect to the original plan?

Manifestation of the Symbols

The kingdom is inside you and it is outside of you.
(*The Gospel of Thomas*)

The Universe had to be created in order to warrant the needed conditions for the appearance of life. These conditions were so rare and specific, that the probability for their formation was incredibly low. Therefore, the overall size of the Universe had to be large enough to provide sufficient statistics to assure the formation of organic life.

The physical world came into being as a result of several increases of the "frequencies" of the oscillations of the field of universal consciousness. These frequencies are superimposed on a carrying wave which covers the entire Universe and determines its physical size. These discrete increases were qualitative in their character. They were needed to accommodate the gradual ascent from the lowest to the highest forms of matter, from the *one-dimensional point* to man. The growth and expansion of the Universe comprised the gradual creation of matter in its variety of forms of stars, galaxies, planets, and the Earth. These discrete increases were like the initial steps of an evolutionary ladder.

According to the scriptures, there were six stages that led to the appearance of man on the Earth. The sequence leading to the appearance of mankind is described in the first Chapter of *Genesis*. The first line of *Genesis* applies to the creation of the Earth:

> In the beginning God created the heaven and the earth.
> (*Genesis 1: 1*)

According to the Big Bang model, the Earth was formed around 4.5 billion years ago, approximately one-third the age of the Universe. It was made by the accumulation of a cloud of dust left over from the formation of the Sun. At that time it was very hot and without an atmosphere. Much of the Earth was molten because of collisions with other bodies which led to extreme volcanism. In the course of time, it acquired an atmosphere from the emission of gases from volcanic outgassing. This early atmosphere contained no oxygen. Mostly, there were gases such as hydrogen sulfide. In *Genesis*, this phase of the creation is referred to as the second stage of the process ("the second day"):

> And God said, Let there be a firmament in the midst of the waters, and let it divide the waters from the waters.
> And God made the firmament, and divided the waters which were under the firmament from the waters which were above the firmament: and it was so.
> And God called the firmament Heaven. And there was evening and there was morning, a second day.
> (*Genesis 1: 6-8*)

Although the Earth seems to be placed far away from the centre of the Universe, in reality it is located within the most subtle nodal region of the Universe. This zone offers all the needed physical conditions for the appearance of organic life. In other words, the nodal region within which the Earth was created was the most sophisticated one among all the regions of the Universe.

As the Earth cooled, it caused the formation of a solid crust. The cooling also led to the formation of clouds. Rains led to the appearance of the oceans:

> And God said, Let the waters under the heaven be gathered together unto one place, and let the dry land appear: and it was so.
> And God called the dry land Earth; and the gathering together of the waters called he Seas.
> (*Genesis 1: 9-10*)

After the formation of the mineral world, the field of consciousness within which the Earth was placed was switched into a higher mode. It formed a set of new nodal regions which contained the most refined and rarefied minerals. Within these regions, the first primitive forms of life appeared. The earliest evidence of life dates from 3.5 billion years ago. These first forms were capable of regulating their nutrition and growth. They were labelled macromolecules. These first forms of life consumed hydrogen sulfide and released oxygen. This gradually changed the atmosphere to the composition that it has today and allowed for the development of other forms of life. This was the beginning of the plants and the vegetation. It corresponds to "the third day" of the process:

> And God said, Let the earth bring forth grass, herb yielding seed, and fruit tree bearing fruit after its kind, wherein is the

> seed thereof, upon the earth: and it was so.
> And the earth brought forth grass, herb yielding seed after its kind, and tree bearing fruit, wherein is the seed thereof, after its kind: and God saw that it was good.
> And there was evening and there was morning, a third day.
> (*Genesis 1, 11-13*)

While the Earth was in this early stage, a giant collision with a planet-sized body formed the Moon:

> And God made two great lights; the greater light to rule the day, and the lesser light to rule the night /…/
> And there was evening and there was morning, a fourth day.
> (*Genesis 1: 16-19*)

When the plants and vegetation were able to grow and multiply, then the next increase in frequency of the field of consciousness took place. This time, the newly formed nodal regions permeated the finest of the vegetal organisms. The effect of this was to produce voluntary sensation and movement, which led to the appearance of the animal world:

> And God said, Let the waters bring forth abundantly the moving creature that hath life, and let fowl fly above the earth in the open firmament of heaven.
> And God created the great sea-monsters, and every living creature that moveth, which the waters brought forth abundantly, after their kinds, and every winged fowl after its

> kind: and God saw that it was good.
> And God blessed them, saying, Be fruitful, and multiply, and fill the waters in the seas, and let fowl multiply in the earth.
> And there was evening and there was morning, a fifth day.
> (*Genesis 1: 20-23*)

There was a massive burst of animal diversity some 500 million years ago when primitive life forms such as the jellyfish, sponges, algae, anemones, worms and arthropods appeared. This period is considered to be the dawn of animal life. It is known as the Cambrian explosion.

After the appearance of the animal world, the increase in frequency of the oscillations of the field of consciousness permeated the finest animal organisms. The result of this permeation was the appearance of mankind with its ordinary faculties of intellect, heart and self. These three faculties constitute a lower version of the original triplicity. It is in this sense that man was made "after our likeness":

> And God said, Let us make man in our image, after our likeness: and let them have dominion over the fish of the sea, and over the fowl of the air, and over the cattle, and over all the earth, and over every creeping thing that creepeth upon the earth.
> And God created man in his own image, in the image of God created he him; male and female created he them.
> (*Genesis 1: 26-27*)

Thus the gradations of these various forms of life were the result of a series of adjustments of the field of universal consciousness. The Earth provided all the necessary conditions for man to appear. The Earth's field of consciousness provided the boundaries for the formation of the nodal regions within which vegetation and animals could appear. In other words, this specific "natural" environment was needed as the base for the appearance and sustenance of the human race. Mankind was the last stage of the creation within the physical world.

The formation of mankind concluded the mechanical phase of the process. According to some mystical references, it led to the simultaneous appearance of a multitude of men in various geographical areas. For example, one of Mohammed's sayings refers to this formation as the appearance of "100,000 Adams."[24] According to science, these men (homo sapiens) appeared between 800,000 and 300,000 years ago. However, the final formation of man's brain took place only some 40,000 – 60,000 years ago. This means that it took over 700,000 years to fill in the specific nodal zone of the field of consciousness that acts as a template for mankind. It was this gradual accommodation to the nodal zone of consciousness that was interpreted by science as man's adaptation to his external environment.

After the appearance of man, "he rested on the seventh day." This part of the process was accomplished:

> And on the seventh day God finished his work which he had made; and he rested on the seventh day from all his work which he had made.
> (*Genesis 2: 2*)

---

[24] *The Meccan Revelations - Volume I*, Ibn Al 'Arabi; edited by Michel Chodkiewicz (Pir Press, New York, 2002, p. 338).

It seemed that everything was done and ready for the transfer to man of the responsibilities for the continuation of the process of creation. Yet, the transfer of responsibility did not take place at that point. Although mankind appeared on the Earth in its physical form, it was not ready yet for its evolutionary function. Mankind appeared on the Earth in complete physical form but with an incomplete mind. Man was equipped with the needed faculties, but these devices were in their latent form. Even if man had been told about them and their use, he was incapable either of activating or employing them correctly. Man, in his natural state, was not prepared yet for his mission. Something else was needed.

Something else had to be done before man could start his participation in the process.

The Human Mind

> The difference between all evolution up to date and the present need for evolution is that for the past ten thousand years or so we have been given the possibility of a conscious evolution. So essential is this more rarefied evolution that our future depends upon it.
> (*Idries Shah*)

The human mind is a reflection of the Macrocosm. Therefore, familiarity with the Macrocosm helps to unfold the mind's inner structure.

Let's recall that the level immediately below the Absolute is described as the Realm. The Realm is the top layer of the Macrocosm; it emanates a blue-print for the evolution of mankind. This blue-print, or matrix, is like the DNA of the human mind. This matrix is to be absorbed, digested and then emulated. The elements of the matrix are encoded onto the field of universal consciousness that cascades across the various strata of the Macrocosm, all the way to the level of ordinary man.

The story of Jacob from *Genesis* may serve as an illustration of this process. Here is a version of this story inserted by Shakespeare in *The Merchant of Venice*:

… mark what Jacob did,
When Laban and himself were compromis'd
That all the eanlings which were streak'd and pied
Should fall as Jacob's hire, the ewes being rank,
In end of autumn turned to the rams,
And when the work of generation was
Between these woolly breeders in the act,
The skilful shepherd peel'd me certain wands,
And in the doing of the deed of kind,
He stuck them up before the fulsome ewes,
Who then conceiving, did in eaning time
Fall parti-colour'd lambs, and those were Jacob's.
(*The Merchant of Venice, I.3*)

In this story Jacob used partially peeled sticks, which he placed in front of breeding ewes. As a result, the ewes gave birth to partly-coloured eanlings (young lambs). According to his contract with Laban, all partly coloured lambs became Jacob's property. We may recognize that an Angel (in the original story Jacob was inspired by an Angel), Jacob, the conceiving ewes, and the young lambs represent the Realm, the world of ideas, the word of symbols, and the physical world, respectively. The peeled sticks represent an evolutionary matrix that is projected from the Realm. In accordance with the Will of the Absolute, this particular matrix was to be implemented within the physical world, i.e., at the level of ordinary man. The matrix (the chequered pattern on the peeled sticks) percolates from the Realm (the Angel) through the world of ideas (Jacob) and the world of symbols (the conceiving ewes) until it reaches the world of ordinary man (the lambs). In this way, the Will of the Absolute may be actualized among ordinary men. It was in this manner that an evolutionary ladder (*Jacob's ladder*) was made available to mankind. By climbing it up, man can ascend from his ordinary, animal-like state, and reach toward the Absolute. Hence, he will be able to fulfil his ultimate purpose.

At every descending stage, the matrix is partially veiled. The veiling is needed to bring the essential elements of the matrix all the way down into the physical world, the world of inferior senses. (The "veiling" of the lower grades of consciousness is referred to by scientists as "the principle of veiled non-locality.") This made the future observer's situation quite a challenging one. He was not only expected to recognize the matrix within his inferior environment. Through incredible personal effort, he needs to activate a set of inner faculties within his mind which will allow him to overcome his natural limitations.

The mode of operation of the Macrocosm indicates that every single process within the physical world takes place according to the state of the field of universal consciousness. The various elemental particles, the vegetal world, the animal world and humankind are all like the constituent organs of a huge cosmic body. They are all subject to a single plan. In this plan, the inner structure of the human mind is a projection of the various strata of the Macrocosm. This is why the human mind may be referred to as the microcosm. This is in accordance with a concept that was laid out by Hermes Trismegistus in the *Emerald Tablet*:

> That which is Below corresponds to that which is Above, and that which is Above, corresponds to that which is Below, to accomplish the miracles of the One Thing.[25]

---

[25] As quoted in a translation by Isaac Newton in *The Chymistry of Isaac Newton* (Keynes MS. 28, King's College Library, Cambridge University).

Over the time of several billions of years, a life process has been developed on Earth and has culminated in the appearance of man. The formation of mankind has been achieved by several ascending modes of oscillations of the field of universal consciousness. At different historical times these various modes of consciousness were activated in succession. Each new mode was higher in its developmental potentiality than the one before. Each oscillation was higher in frequency than the previous one. These various modes of oscillations may be arbitrarily referred to as *constructive*, *vital*, *automatic*, and *rational*. As they were switched-in in turn, they gave rise to the progression from molecule to man. The birth of each human consists of a compressed sequence of this progression.

The first mode of these oscillations, *constructive*, led to the formation of the mineral form: man was conceived from a blood clot. Afterwards, the *vital* mode transformed that initial mineral form into the vegetal form: in his mother's womb he was capable of regulating his nutrition and growth. Then the *automatic* mode followed: after birth, he experienced the animal form by acquiring voluntary sensation and movement.

Jalaluddin Rumi, a 13th century Persian poet, described this process in the following way:

> First, man appeared in the class of inorganic things. Next he passed from the inorganic class into that of plants. For some time he lived as one of the plants, remembering nothing of his previous inorganic state. Then he passed from the vegetal to the animal state. He had no remembrance of his state as a plant, except the inclination he felt towards plants. Especially at the time of spring, he felt an attraction to green trees and the aroma of flowers. Then again, man was led from the animal state towards humanity. In this way he did advance from one world of being to another, till he became intelligent and

> rational. He has no memory of his former states. During these periods man did not know where he was going but he was being taken on a long journey nonetheless.
> From his present state, man needs to continue his migration so that he may escape from his rationality and intellectuality which are driven mostly by greed and egotism.
> There are a hundred thousand more marvellous states ahead of him. He fell asleep and became oblivious of the past. This world is the sleeper's dream and the sleeper's fancies. Till all of a sudden there shall rise the dawn of death and he shall be delivered from ignorance.
> (*Mathnawi, Book IV, 3637-55*)

The *rational* mode led to the appearance of the ordinary layer of the human mind, the rational mind. This is an earthly component, a natural one. Its operation is limited to the physical senses. It is acquired by the body with the body's first breath. Shakespeare's King Lear describes this in the following way:

> When we are born, we cry that we are come
> To this great stage of fools.
> (*King Lear, IV.6*)

The *rational* mind may be thought of as a fine substance that possesses the capacity for nutrition, growth and sense-perception. It is this part of the mind that is the source of the body's life. Consequently, when divorced from the body, it causes its death. (When separated from the body, the rational mind is often referred to as the "natural soul.") When the body dies and disintegrates, the rational mind goes through a similar process of dissolution. It takes some time for this subtle substance to dissipate after its separation from the body. Ultimately, it disappears to nothingness.

The rational mind consists of man's ordinary faculties of *intellect*, *heart*, and *self*. These three faculties govern man's physical, emotional, and intellectual life. In the hierarchical structure of the mind, the ordinary faculties constitute the lowest level, i.e., the natural state of mind. The ordinary faculties are man's lowest form of the original triplicity.

The natural or rational mind is focussed on man's survival as well as on his desires for and pursuit of pleasure, ambition, and self-importance. All of these are the objectives of the *self-faculty*. This is why this faculty is also referred to as the ego-self.

The *heart faculty* entertains desires and emotions such as feelings of sensual attachments and hatred, showing bravery or cowardice, forming an intention and carrying out a particular action.

The qualities that are attributed to the *intellect faculty* are understanding and knowledge, the capacity to perceive, to recollect the things of the past and plan for the things of the future.

For methodological purposes, the position of the ordinary faculties within the human body has been arbitrarily determined.[26] The self-faculty permeates the whole body, but its nodal zone is firmly rooted in the liver; the heart faculty is present throughout the whole body, but is firmly rooted in the physical heart; and the intellect faculty also pervades the entire body, but it is firmly rooted in the brain. Shakespeare's Duke Orsino alludes to this structure in the following lines:

---

[26] *The Sacred Knowledge*, Shah Waliullah (The Octagon Press, London, 1982, p. 16).

How will she love, when the rich golden shaft
Hath kill'd the flock of all affections else
That live in her. When liver, brain and heart,
These sovereign thrones, are all supplied, and fill'd
Her sweet perfections with one self king.
(*Twelfth Night, I.1*)

The sum-total of the relationships between these three faculties forms an individual's character and personality.

In its ordinary state, the rational mind is dominated by the self-faculty. The self-faculty subdues the heart faculty and the intellect faculty. Both of them are employed to satisfy the ego-self. Such behaviour undermines and contaminates the proper functioning of the intellect and the heart. In such a state, man is not able to make full use of his potential. The natural mind is developmentally sterile. It is incapable of playing an active part in the creation. In their natural state, the faculties are a degenerated form of the original triplicity.

Within the overall structure, the ordinary faculties of self, heart and intellect form the upper limit of consciousness of the physical world; the rational mind operates entirely within the corporeal world. The *rational mind* was the last and highest mode of oscillation activated within space-time. It was the last stage of the mechanical creation. With its appearance, the process of mechanical creation was completed. It was at this point that, as stated in *Genesis*, the Absolute "rested":

> He rested on the seventh day from all his work which he had made.
> (*Genesis 2: 2*)

In order to activate his dormant evolutionary potentialities, man needs to be exposed to the modes of oscillations of the field of consciousness which are available within the Macrocosm. Only through putting skillfully his mind into resonance with these higher modes may he activate or "awaken" his inner potentialities. But, in his natural state, he is not even capable of reaching the buffer zone between the visible and the invisible; he is not even aware of his potentiality. How, then, can he be expected to get out of this seemingly impossible situation?

This situation is the result of granting mankind free-will. It is up to man to try and make efforts. In this way he may escape extinction and prolong his existence. But this is not evolution as it is commonly understood. This evolution is deliberate; it requires man's conscious participation in the creative process. It is his choice.

First of all, man has to prepare himself for such a "journey."

The first requirement is that the inner hierarchy of his rational mind be rearranged. The re-arrangement is a two-step process. The first step is achieved when the intellect faculty gains control over the other two faculties. The second step requires that the inner relationships should be rearranged in such a way that the intellect should control the heart faculty, and the heart faculty should rule over the ego-self. From the combination of these two degrees of control, further developmental stages may result.

An ancient parable may be used to explain the function of the proper alignment of the ordinary faculties. In this parable the inner structure of the mind is compared to a chariot:

> A driver is seated in a chariot that is propelled by a horse. The chariot represents the self-faculty, i.e., the outward form which allows the driver to move toward its objective. The horse, which is the motive power that enables an intention to be actualized and a particular action to be carried out, represents the heart faculty. The driver represents the intellect faculty. It is the intellect faculty that, in a superior manner, perceives the purpose and possibility of the situation and makes it possible for the chariot to move forward and achieve its objective. One of the three, on its own, will be able to fulfill its limited function. However, the combined function of reaching its destination cannot be realized unless all three faculties are aligned in the right way.[27]

We may recognize that such an arrangement of the intellect, the heart, and the self is an earthly reflection of the original triplicity. It is this sort of "chariot" that may take man on a cosmic journey. This is the first step in the preparation for the journey through cosmic consciousness. It is this first step that is the focus and the objective of conventional religions. Religions offer various prescriptions and practices for subduing the ego-self.

---

[27] This version is based on "The Chariot," a parable included in *Tales of the Dervishes*, Idries Shah (The Octagon Press, London, 1982, p. 207).

***

Unlike the ego-self, the intellect and the heart are not homogeneous. The intellect faculty and the heart faculty consist of a multi-layered inner structure. These inner layers are called the inner faculties of the mind. In the natural state of the mind, these inner faculties remain in their latent forms. They may be activated when they are in resonance with specific higher modes of the oscillations of the field of universal consciousness. Therefore, the process of the evolution of the human mind may be described as an interaction between the ordinary faculties and the higher modes of the field of universal consciousness. When the ordinary faculties are in resonance with the macrocosmic oscillations, they are split. In this way their finer inner structure is unfolded. This may be compared to the splitting of atomic orbitals. In their natural form, atomic orbitals remain degenerate. When exposed to a strong external magnetic field, they are split into several sub-levels, in accordance with their characteristic wavefunctions.

In the case of the human mind, the higher modes of the field of consciousness correspond to the intense magnetic fields of the above analogy. These higher modes, however, do not operate in space-time. They are available only within the various zones of the Macrocosm, i.e., above the buffer zone. This makes man's situation rather a challenging one. But let's leave him here for a while as he is contemplating his choices. Right now, he is *asleep* to his evolutionary potential; he is not ready yet to embark on his journey. At this point, his situation corresponds to that of Hamlet, who also meditates on his options, "to be or not to be." Hamlet's "to be" means "to die, to sleep no more."[28] In this context, "to die" implies to *wake-up from sleep* by taking "arms against a sea of troubles":

[28] Some editors of Shakespeare's plays change the punctuation from "to die, to sleep" into "to die: to sleep," or "to die – to sleep," or "to die, to sleep –."

> To be, or not to be, that is the question:
> Whether 'tis nobler in the mind to suffer
> The slings and arrows of outrageous fortune,
> Or to take arms against a sea of troubles,
> And by opposing end them: to die, to sleep
> No more; …
> (*Hamlet, III.1*)

In the meantime, let's explore the inner world that is hidden within the deeper layers of the human mind.

***

The unfolding of the inner layers of the mind is a multi-step process. At each step a subtler layer of the intellect or the heart faculty is activated. The process of the activation of these inner layers of the mind constitutes man's journey through the various strata of cosmic consciousness.

The journey through cosmic consciousness may be initiated either through the unfolding of the subtle layers of the intellect faculty (the path of the intellect) or through unfolding the subtle layers of the heart faculty (the path of the heart). The choice is determined by a person's initial predisposition.

The first subtle layer is veiled within the heart faculty. This layer may be activated when the mind is exposed to the field of consciousness within the world of symbols. This is the first step of the ascent into the Macrocosm. This first layer is called the Spirit.[29] The ordinary faculty of the heart is driven by earthly attachments and longing. The Spirit may be activated when the heart faculty is detached from these earthly bonds. Through the Spirit, the traveler is capable of forming a supreme desire. This makes everything else of secondary importance. Because of this inclination towards the supreme objective, the Spirit overrides all sorts of earthly indulgencies driven by the ego-self. It brings the traveler closer to his ultimate objective; it allows experiencing closeness with it. When activated, the Spirit becomes a super-corporeal part of the mind.

The subtle layer of the intellect faculty is called the Secret. The Secret appears when the intellect is freed from its earthly inclinations. The Secret manifests itself as an intuitive certainty about certain episodes and occurrences, without knowing how one has arrived at it. Through this faculty, the mind is occasionally flooded with images and visions. It is this faculty that allows foreseeing and even influencing certain future events. Through its operation, the traveler may also be able to see into the minds of others and gain access to unspoken thoughts. It is at this point that "belief" melts into "knowledge." When something is experienced, then there is no need for belief.

Similarly to the Spirit, the Secret is super-corporeal. This faculty of the mind and its experiences are not comprehensible to the ordinary intellect. The intellect has an area within which it may wander about and exert itself, but beyond that area it cannot pass; it has no access to the conditions existing outside its space-time limited perception. If the intellect moves beyond the range of what it can perceive, it becomes confused. Of course, the intellect does

---

[29] The terms used here to designate the inner faculties (Spirit, Secret, Mysterious, Concealed, and Supracognitive) are arbitrary. Various authors use different terms. What is important, however, is their relationship with the ordinary faculties, their inner hierarchy, and the forms of their manifestations.

not perceive the situation like this; in fact, it emphatically denies that it is so.

The Spirit and the Secret form the *creative mind.* Sometimes, the creative mind is referred to as the creative soul. The creative mind operates within the various levels of the world of symbols. In the natural mind, the Spirit and the Secret are veiled. Their veiling is imposed by the inner buffer zone within the mind that separates the physical world from the invisible worlds. This means that through the activation of the Spirit and the Secret, man is able to pass through the buffer zone and enter into the lower levels of the world of symbols. The Spirit and the Secret are the first experiences for a traveller who overcomes the limitations imposed by corporality. As emphasized earlier, these subtle layers are a finer layer of consciousness that is interwoven into the natural human mind.

After experiencing the various sublayers of the Spirit and the Secret, the traveler is ready to step into the world of ideas. Then, he may be exposed to higher modes of oscillations of the field of universal consciousness. This time, such an exposure leads to the activation of the next layer of the inner structure of the mind. This deeper layer of the mind is called the *sublime mind* or the sublime soul. The sublime mind consists of the hidden faculties. The hidden faculties allow the traveler to explore the various aspects of the world of ideas. By exploring the various spheres of the world of ideas, one is gradually cleansing his mind from remaining earthly residues.

There are innumerable sublayers within the world of ideas and the world of symbols. Sometimes the symbolic number "ninety nine" is used to denote the various names of the inner layers. This symbolic representation indicates that one has to experience "ninety nine" states. Consequently, the number "one hundred" indicates the completion of this particular stage of the journey.

While in the world of ideas, the mind is a composite mix of the *rational*, the *creative* and the *sublime* parts. For every traveler, there will be a different ratio of these composite parts. Depending on the relative strength of the creative and sublime parts, the traveler will be exposed to different experiences.

At one point the traveler may be able to perceive the presence of a shining spot that is deeply hidden in the midst of his mind. This is the traveler's first encounter with the "dot" of the mystical formula. This "dot" is the projection of *Pure Intellect*; a sample of the Absolute left within man's mind. It is then that one starts to perceive the presence of the Prime Architect. It is the "dot" that makes man to be "in our image":

> Let us make man in our image, after our likeness.
> (*Genesis 1: 26*)

It is a characteristic of the Absolute that, at one stage, it is entirely engrossed in self-contemplation; while at other stages, despite its purity, it descends into the deepest level of the human mind. However, in the course of that descent it loses none of its purity.

So, when a traveler plunges into deep contemplation, then at the utmost limit of his vision is that essential shining "dot." He imagines that this point is in the middle of his own inner being; whereas in fact the "dot" dwells in a glorious abode.[30] As the traveler continues his journey towards the Realm, the "dot" becomes much brighter. Its brightness depends on the purity of the mind. If there are still some earthly traces, the traveler will be able to perceive the presence of the "dot," but he will not be able to

---

[30] *The Sacred Knowledge*, Shah Waliullah, p. 69 (see Note #26).

envisage it. This corresponds to a situation when the creative part of the mind is dominant. If the traveler follows the path of the heart, the presence of the "dot" will lead to the activation of the hidden faculty that is called the Mysterious. The Mysterious faculty is characterized by indescribable peace and tranquility. The traveler who experiences it becomes liberated from desires and dependence. He discovers the uselessness of indulging in ecstatic experiences associated with the Spirit. Indulgence in ecstasy involves the experiences of emotional raptures which may intoxicate but do not raise the traveler to a higher state. The traveler must overcome this attraction in order to detach himself completely from the physical world. Through the Mysterious faculty, the traveler may gain permanence within the world of ideas.

If the traveler follows the path of the intellect, the presence of the "dot" will lead to the activation of another hidden faculty which is called the Concealed. The Concealed faculty allows one to contemplate the world of ideas, to comprehend it, to be present before it, and to gain deep knowledge of it. There is a difference between the contemplation experienced by the Concealed faculty and the certitude which flows into the Secret faculty. The difference is that contemplation by the Concealed faculty takes place in the presence of the thing sought; while the certitude of the Secret faculty means being certain about things absent and acknowledging the unseen. The Concealed faculty may lead to the annihilation of the traveler within the world of ideas. He gains cosmic permanence.

Then there is another possibility. It happens when the hidden and the subtle faculties are arranged together according to a specific pattern. It is then that all the four inner faculties of the Spirit, the Secret, the Mysterious and the Concealed may operate harmoniously in a united and balanced manner. When arranged in accordance to that specific pattern, they may be put in resonance with the oscillation of the Realm. When in resonance with the

Realm, a new faculty of perception appears in their midst. This fifth faculty is called the Supracognitive:

> For there are five trees in paradise for you; they do not change, summer or winter, and their leaves do not fall. Whoever knows them will not taste death.
> (*The Gospel of Thomas, 19*)

At one point, the sublime mind may be fully separated from the creative mind. Then it is possible to have the Mysterious faculty and the Concealed faculty in perfect harmony with the Supracognitive faculty. In that state all are perfectly balanced; none of them dominates the others. The three of them form the highest form of the ascending triplicity.

At the highest level of its manifestation, the Supracognitive faculty may rise up in agitation and tear down the remaining traces of the veil. It may then completely dominate the entire mind. The Supracognitive faculty is transmuted into Pure Intellect. In this state everything else is annihilated and only Pure Intellect remains. At this point, the ascending triplicity coincides with the original triplicity; it becomes part of the Realm. It is this triplicity that Jesus referred to as "Father" (Pure Essence), "Holy Spirit" (the highest mode of the field of universal consciousness), and "Son" (Pure Intellect). It is then that the divine synthesis is accomplished. The traveller is annihilated in the "dot." He becomes a part of the original triplicity; he is absorbed within the Realm. While in that state, the traveler gains complete knowledge and understanding. He arrives at the stage of completion and perfection. He is united with the Absolute. The ascending loop of the creation initiated with the Big Bang is completed. This state is beyond any description or comprehension. This experience belongs to the fully realized Man.

The journey through the various subtle and hidden faculties has been symbolically illustrated by Fariduddin Attar, a 12th century Persian poet, in his tale entitled *The Conference of the Birds.* In this tale, the birds are called together by the hoopoe, their guide. The hoopoe proposes that the birds should start on a quest to find their mysterious King. The King is called Simurgh, and he lives in the Mountains of Kaf. The hoopoe tells the birds that in their quest they have to traverse seven valleys. Each of these valleys represents one of the states associated with the subtle or hidden faculties:

- The first valley is the Valley of the Quest (*the re-arrangement of the ordinary faculties*).
- The second valley is the Valley of Love (*the activation of the Spirit*).
- The third valley is the Valley of Intuitive Knowledge (*the activation of the Secret*).
- The fourth valley is the Valley of Detachment (*the activation of the Mysterious*).
- The fifth valley is the Valley of Unity (*the activation of the Concealed*).
- The sixth valley is the Valley of Astonishment (*the activation of the Supracognitive*).
- The seventh valley is the Valley of Death (*annihilation in Pure Essence*).

After passing through the seven valleys, the group of birds arrives at their ultimate destination. It is there that the birds meet Simurgh:

> So then, out of all those thousands of birds, only thirty reached the end of the journey. And even these were

bewildered, weary and dejected, with neither feathers nor wings. But now they were at the door of the Majesty that cannot be described, whose essence is incomprehensible – the Being who is beyond human reason and knowledge. Then flashed the lightning of fulfilment, and a hundred worlds were consumed in a moment. They saw thousands of suns each more resplendent than the other, thousands of moons and stars all equally beautiful, and seeing all this they were amazed and agitated like a dancing atom of dust, and they cried out: 'O Thou who art more radiant than the sun! Thou, who hast reduced the sun to an atom, how can we appear before Thee? Ah, why have we so uselessly endured all this suffering on the Way? Having renounced ourselves and all things, we now cannot obtain that for which we have striven. Here, it little matters whether we exist or not.'

Then the birds, who were so disheartened, sank into despair. A long time passed. When, at a propitious moment, the door suddenly opened, there stepped out a noble chamberlain, one of the courtiers of the Supreme Majesty. He looked them over and saw that out of thousands only these thirty birds were left.

He said: 'Now then, O Birds, where have you come from, and what are you doing here? What is your name? O you who are destitute of everything, where is your home? What do they call you in the world? What can be done with a feeble handful of dust like you?'

'We have come,' they said, 'to acknowledge the Simurgh as our king. Through love and desire for him we have lost our reason and our peace of mind. Very long ago, when we started on this journey, we were thousands, and now only thirty of us have arrived at this sublime court.' ...

Then the Chamberlain, having tested them, opened the door: and as he drew aside a hundred curtains, one after the other, a new world beyond the veil was revealed. Now was the light of lights manifested, and all of them sat down on the throne, the seat of the Majesty and Glory. They were given a writing which they were told to read through; and reading this, and pondering, they were able to understand their state. When they were completely at peace and detached from all things, they became aware that the Simurgh was there with them, and a new life begun for them in the Simurgh. All that they had done previously was washed away. The sun of majesty sent forth his rays, and in the reflection of each other's faces, these thirty birds of the outer world contemplated the face of the Simurgh of the inner world. This so astonished them that they did not know if they were still themselves or if they had become the Simurgh. At last, in a state of contemplation, they realized that they were the Simurgh and that the Simurgh was the thirty birds. When they gazed at the Simurgh they saw that it was truly the Simurgh who was there, and when they turned their eyes towards themselves they saw that they themselves were the Simurgh. And perceiving both at once, themselves and Him, they realized that they and the Simurgh were one and the same being. No one in the world has ever heard of anything equal to it.

Then they gave themselves up to meditation, and after a little they asked the Simurgh, without the use of tongues, to reveal to them the secret of the mystery of the unity and plurality of beings. The Simurgh, also without speaking, made this reply: 'The sun of my majesty is a mirror. He who sees himself therein sees his soul and his body, and sees them completely. Since you have come as thirty birds, you will see thirty birds in this mirror. If forty or fifty were to come, it would be the same. Although you are not completely changed you see yourselves as you were before.' …

'All that you have known, all that you have seen, all that you have said or heard – all this is no longer that. When you crossed the valleys of the Spiritual Way and when you performed good tasks, you did all this by my actions; and you were able to see the valleys of my essence and my perfections. You, who are only thirty birds, did well to be astonished,

> impatient and wondering. But I am more than thirty birds. I am the very essence of the true Simurgh. Annihilate then yourselves gloriously and joyfully in me, and in me you shall find yourselves.'
>
> Thereupon, the birds at last lost themselves for ever in the Simurgh – the shadow was lost in the sun, and that is all.[31]

The Simurgh, which means "thirty birds," is a code phrase which in the abjad system means "the development of the mind through China." In both Persian and Arabic, "China" stands for the concealed concept of meditation and the developmental methodology.[32] This is the origin of the saying "Seek knowledge, even unto China." In this context, the word "thirty" is an equivalent to the highest state of consciousness, symbolically represented by the original triplicity. It is in this way that man retraces the various stages of the Macrocosm and ascends towards the Absolute; he returns to his origin. It is then that the divine wish "I wished to be known" is fulfilled. The loop of creative synthesis is closed.

The human mind at this stage is called the *supracognitive mind.* Sometimes the term "angelic soul" is used to mark this particular stage of the human evolution. While in this state, the traveler is able to see things as they really are, to understand the affinity and unity of seemingly different things, and to perceive the role of man. Then the traveler may discover that the entire Macrocosm is reproduced within his mind.

---

[31] *The Conference of the Birds*, Fariduddin Attar; translated by S.C. Nott (Continuum Publishing, New York, 2000, p. 145-148).
[32] *The Sufis*, Idries Shah, p. 395 (see Note #13).

***

There may be differences in understanding of these various experiences because it is not within the capacity of every traveler to recognize them correctly. Some travelers may conclude that they experienced consciousness at the level of the Concealed faculty, while in reality they were exposed to the certitude of the Secret faculty. Or a traveler mind's may be overflowed by experiencing consciousness of the Supracognitive faculty, but may believe that he experienced unity with the Absolute.

If those, who have not been prepared correctly, attempt to venture into the world of symbols, they will be disturbed by such an experience. The following story entitled "Paradise of Song" is an allegorical illustration of an experience that was driven by an incorrect intention ("It would not be right, I know that") and undertaken without guidance:

> Ahangar was a mighty swordsmith who lived in one of Afghanistan's remote eastern valleys. In time of peace he made steel ploughs, horse shoes, and above all, he sang.
>
> The songs of Ahangar, who is known by different names in various parts of Central Asia, were eagerly listened to by the people of the valleys. They came from the forests of giant walnut-trees, from the snow-capped Hindu-Kush, from Qataghan and Badakhshan, from Khanabad and Kunar, from Herat and Paghman, to hear his songs.
>
> Above all, the people came to hear the song of all songs, which was Ahangar's Song of the Valley of Paradise.
>
> This song had a haunting quality, and a strange lilt, and most of all it had a story which was so strange that people felt they knew the remote Valley of Paradise. Often they asked him to sing it when he was not in the mood to do so, and he would refuse. Sometimes people asked him whether the Valley of

Paradise was truly real, and Ahangar could only say:

'The Valley of the Song is as real as real can be.'

'But how do you know?' the people would ask, 'Have you ever been there?'

'Not in any ordinary way,' said Ahangar.

To Ahangar, and nearly all the people who heard him, the Valley of the Song was, however, real as real can be.

Aisha, a local maiden whom he loved, doubted whether there was such a place. So, too, did Hasan, a braggart and fearsome swordsman who swore to marry Aisha, and who lost no opportunity of laughing at the smith.

One day, when the villagers were sitting around silently after Ahangar had been telling his tale to them, Hasan spoke:

'If you believe that this valley is so real and that it is, as you say, in those mountains of Sangan yonder, where the blue haze rises, why do you not try to find it?'

'It would not be right, I know that,' said Ahangar.

'You know what it is convenient to know, and do not know what you do not want to know!' shouted Hasan. 'Now, my friend, I propose a test. You love Aisha, but she does not trust you. She has no faith in this absurd Valley of yours. You could never marry her, because when there is no confidence between man and wife, they are not happy and all manner of evils result.'

'Do you expect me to go to the valley, then?' asked Ahangar.

'Yes,' said Hasan and all the audience together.

'If I go and return safely, will Aisha consent to marry me?' asked Ahangar.

'Yes,' murmured Aisha.

So Ahangar, collecting some dried mulberries and a scrap of bread, set off for the distant mountains.

He climbed and climbed, until he came to a wall which encircled the entire range. When he had ascended its sheer sides, there was another wall, even more precipitous than the first. After that there was a third, then a fourth, and finally a fifth wall.

Descending on the other side, Ahangar found that he was in a valley, strikingly similar to his own.

People came out to welcome him, and as he saw them, Ahangar realized that something very strange was happening.

Months later, Ahangar the Smith, walking like an old man, limped into his native village, and made for his humble hut. As word of his return spread throughout the countryside, people gathered in front of his home to hear what his adventures had been.

Hasan the swordsman spoke for them all, and called Ahangar to his window.

There was a gasp as everyone saw how old he had become.

'Well, Master Ahangar, and did you reach the Valley of Paradise?'

'I did.'

'And what was it like?'

Ahangar, fumbling for his words, looked at the assembled people with a weariness and hopelessness that he had never felt before. He said:

'I climbed and I climbed, and I climbed. When it seemed as though there could be no human habitation in such a desolate place, and after many trials and disappointments, I came upon a valley. This valley was exactly like the one in which we live. And then I saw the people. Those people are not only like us people: they are *the same people*. For every Hasan, every Aisha, every Ahangar, every anybody whom we have here, there is another one, exactly the same in that valley.

'These are likeness and reflections to us, when we see such things. But it is we who are the likeness and reflection of them - we who are here, we are their twins …'

Everybody thought that Ahangar had gone mad through his privations, and Aisha married Hasan the swordsman. Ahangar rapidly grew old and died. And all the people, everyone who had heard this story from the lips of Ahangar, first lost heart in

> their lives, then grew old and died, for they felt that something was going to happen over which they had no control and from which they had no hope, and so they lost interest in life itself.
>
> It is only once in a thousand years that this secret is seen by man. When he sees it, he is changed. When he tells its bare facts to others, they wither and die out.
>
> People think that such an event is a catastrophe, and so they must not know about it, for they cannot understand (such is the nature of their ordinary life) that they have more selves than one, more hopes than one, more chance than one - up there, in the Paradise of the Song of Ahangar the mighty smith.[33]

It is also possible that certain experiences may fully saturate the capacity of a particular person. In that ecstatic state, this person's mind will be completely annihilated within one of the above described higher states. This was the case of Bayazid of Bistam who lived in the 9th century:

> While being in an ecstatic state, Bayazid uttered the words "Glory onto me." His disciples protested, "It is not appropriate for you to say that." Bayazid told them, "Friends, beware! If you are sincere men, then when I utter these words again, take your knives and swords and strike me, so you can be among those who receive the approval of God." When that same state returned to Bayazid, he started to exclaim: "Glory onto me! How very great is my Glory." Some of the disciples pulled out their knives and attacked him. When they recovered their senses, they realized that they had cut their own hands,

---

[33] "Paradise of Song" included in *Wisdom of the Idiots*, Idries Shah (The Octagon Press, London, 1971, p. 77.)

> and wounded their bellies and chests. But the others, who did not strike, were unharmed and neither was Bayazid.[34]

This is why, for example, Thomas would not reveal his state of consciousness to other disciples of Jesus:

> Jesus said to his followers, "Compare me to something and tell me what I am like."
> Simon Peter said to him, "You are like a righteous messenger."
> Matthew said to him, "You are like a wise philosopher."
> Thomas said to him, "Teacher, my mouth is utterly unable to say what you are like."
> Jesus said, "I am not your teacher. Because you have drunk, you have become intoxicated from the bubbling spring that I have tended."
> And he took him, and withdrew, and spoke three sayings to him.
> When Thomas came back to his friends, they asked him, "What did Jesus say to you?"
> Thomas said to them, "If I tell you one of the sayings he spoke to me, you will pick up rocks and stone me, and fire will come from the rocks and consume you."
> (*The Gospel of Thomas, 13*)

Exposure to higher modes of consciousness may be challenging when the newly activated subtle faculties are still dominated by egotistic tendencies. Humans must purify, at least partially, their

---

[34] *Maître et Disciple*, Sultan Valad; translated into French by Eva de Vitray-Meyerovitch (Éditions Sindbad, Paris, 1982). English translation is from *Elucidation* (Troubadour Publications, to be published).

ego-self before they can arrive at higher zones of consciousness. Any attempt to activate the subtle faculties within an unregenerate personality will end up with an aberration. In such cases, exposure to heightened states of consciousness may not only provide access to new "powers" but it will also boost up destructive tendencies. When this happens, the individual may gain a deepening in intuitive knowledge corresponding to the faculty involved. But if this is not part of a comprehensive development, the mind will try, vainly, to equilibrate itself around this hypertrophy. The consequences include one-sided mental phenomena, exaggerated ideas of self-importance, the surfacing of undesirable qualities or a mental deterioration. Instead of rising up to higher spheres of functioning, such an exposure corrupts, strengthens the ego-self and reduces man down to a kind of bestial life.

It is this mechanism that is responsible for the appearance in the literature of such fictitious creatures as devils, demons, witches, etc. The possibility for the appearance of malignant minds is necessitated by the measure of freewill which man has and its consequences may not be annulled by the macrocosmic forces, no matter how much is at stake. All that may be done is to contrive life situations which will provide increased opportunities for man to choose differently. For example, Shakespeare used the witches in *Macbeth* to illustrate the modus operandi of such enhanced but destabilizing tendencies:

> The enhanced negative tendencies are symbolically represented by Hecate and her witches. Macbeth is their target. Hecate exercises her influence by playing on Macbeth's unbalanced personality, i.e., his complexes and insecurity. At first, Hecate is angry at the witches, because they made a mistake in investing their influence in a rather weak character. (In this way Shakespeare indicates that there is no such thing as absolute

> evil; the witches can only amplify existing negative tendencies.) From Hecate's perspective, Macbeth is not vicious enough; he is not able to serve evil efficiently enough. As a matter of fact, he is rather weak and shaky in handling himself. Macbeth's "loves for his own ends, not for you" may spoil Hecate's overall objective. Therefore, she tells the witches that she will arrange for Macbeth to come back to them looking for further prophecies. And when he comes, they must summon visions and spirits whose messages will enhance Macbeth's viciousness by filling him with a false sense of security, because "security is mortals' chiefest enemy." The witches' prophecies are based on a template that consists of an accurate presentation of future events. But their visions are projected in a distorted manner and with a deceptive emphasis. It is in this manner that the witches' prophecies become self-fulfilling. Under the witches' influence, Macbeth is turned into a brutal, murderous villain.[35]

It was this sort of enhanced destabilizing tendencies that interfered with the evolutionary process in ancient times. At that time, partially developed men abandoned their evolutionary responsibilities. It seems that their creative minds were activated prematurely. Instead of sustaining the process, these men allowed themselves to follow their still raw egos. They took advantage of their superior skills in order to satisfy their egoistic desires. As a result, the ancient world was temporarily disconnected from the evolutionary process. A break occurred between the Realm and the world of symbols; the world of symbols became contaminated. This contamination was manifested by the appearance of various "demigods" who interfered in human affairs. There is a reference in *Genesis* to these events:

---

[35] Extracted by the author from *Shakespeare's Elephant in Darkness England*, W. Jamroz (Troubadour Publications, Montreal, 2016, p. 90).

> That the sons of God saw the daughters of men that they were fair; and took them wives of all that they chose. …
> And the Lord saw that the wickedness of man was great in the earth, and that every imagination of the thoughts of his heart was only evil continually.
> (*Genesis 6: 2-5*)

This degenerated situation was recorded in ancient myths and legends. The ancient myths are a record of an evolutionary disruption that was caused by an event that occurred in antiquity. As a result of this ancient event, humanity ended-up in evolutionary chaos. The events in Greek and Roman mythologies are symbolic illustrations of the consequences of that evolutionary break. Driven by their selfishness and sensuality, these various demigods and demigoddesses abused their extraordinary powers and the responsibility that they had been charged with. Instead of overseeing the evolutionary process, these ancient men and women focused their activities on pursuing inferior objectives. This led to a corruption of the process. Mankind was separated from the Realm. The above Biblical reference was inserted just before the episode with Noah. It may be presumed, therefore, that the Biblical "flood" was a form of cleansing operation that was needed to bring mankind back onto the right evolutionary track.

***

For methodological and illustrative purposes, the inner faculties of the mind are associated with specific colours. The colours are used as a symbolic representation of the various modes of oscillations of the field of universal consciousness. In this way, colours may be used as a developmental tool.[36] The colours are chosen in such a way that their subtractive properties reflect the hierarchical relationship between the inner faculties. Namely, there is a hierarchy among the colours, i.e., yellow and red are considered to be the lowest colours. They appear first, when light (white) is subtracted from blackness. Consequently, white and black are higher (prime) colours; and green is the highest because it is the last colour to appear when the white is being removed from the black.[37] Accordingly, the colours yellow and red are used to denote the inner faculties of the heart (the Spirit and the Mysterious); the colors white and black are chosen for the inner faculties of the intellect (the Secret and the Concealed). The color green is indicative of the Supracognitive faculty.

This colour code was the origin of the "halo" used in religious arts. A halo is a crown of light rays that surrounds a person. It was used to indicate holy or sacred figures. It was usually yellow (golden) to mark the activation of the Spirit.

In addition to colours, the hierarchical relationship between the various modes of oscillations may also be marked in literary descriptions by "age" or by "height." In this convention, "younger" or "taller" means a higher mode. This type of code is used to illustrate the various stages of the development of the human mind. Some mystical authors use female characters to represent the various modes of oscillations. The inner faculties in their latent states are represented by young men. A successful activation of one of the inner faculties is marked as a marriage.

Shakespeare's plays are one of the most precise illustrations of the most recent phase of the evolution of the human mind. The plays

---

[36] The selection of colours is specific to the methodology of the activation of the inner faculties. This selection is at the discretion of a spiritual mentor who oversees the process.

[37] *Goethe's Scientific Consciousness*, Henri Bortoft (The Institute for Cultural Research, Turnbridge Wells, England, 1986, p. 11).

form a narrative that illustrates the development of the human mind in the context of the western civilization.[38] Shakespeare is very consistent in identifying the evolutionary function of his heroines by indicating their relative age, their relative height, or the colour of their hair, clothes or complexion. For example, Shakespeare refers to the symbolism of the colour code in this seemingly meaningless exchange in *Love's Labour's Lost* between Don Adriano and Moth, his Page:

Don Adriano:
"Who was Samson's love, my dear Moth?"
Moth:
"A Woman, Master."
Don Adriano:
"Of what complexion?"
Moth:
"Of all the four, or the three, or the two, or one of the four."
Don Adriano:
"Tell me precisely of what complexion."
Moth:
"Of the sea-water Green, sir."
Don Adriano:
"Is that one of the four complexions?"
Moth:
"As I have read, sir; and the best of them too."
Don Adriano:
"Green indeed is the colour of Lovers;"
(*Love's Labour's Lost, I.2*)

---

[38] *Shakespeare's Elephant in Darkness England*, W. Jamroz, p. 108 (see Note #35).

***

In the search for the ultimate elementary particles, scientists discovered the inner structure of matter. The discovery has been acclaimed as one of the greatest successes of elementary particles physics. And it turns out that the inner structure of matter and that of the human mind are somewhat similar. As there is no room for "soul" in science, obviously the physicists would not use the various levels of consciousness to describe the inner layers of matter. Yet, the mystical description of the mind's structure has found its reflection within the hard core of quantum physics.

Up to the 1960s, it was believed that protons and neutrons were the elementary particles of matter. Then, in 1968, physicists discovered that the proton contains an inner structure. The experiment carried out at the Stanford Linear Accelerator Center (SLAC) showed that the proton's structure is not homogeneous. In this experiment, protons were accelerated to high speeds and then they were made to collide with other protons or electrons. As the result, the protons were smashed into a number of smaller particles. These smaller particles became known as quarks.

Quarks are the only elementary particles in the Standard Model of particle physics to experience all four fundamental forces. Quarks, however, are never directly observed or found in isolation. They cannot be isolated and, therefore, cannot be directly observed in normal conditions. This means that one cannot have a single quark on its own.

Upon closer inspection, it has been determined that each proton and neutron is built, coincidently, from a triplicity of quarks. In other words, the basic arrangement of matter is in a form of triplicity, a remote echo of the original cosmic triplicity. Just like the mystics, the physicists have chosen a colour code to indicate the mutual compatibility between the various quarks. However, instead of the subtractive (hierarchical) model, the physicists use the additive model for the colour coding. The additive model is non-hierarchical; all colours are qualitatively equal to each other. The physicists have arbitrarily labeled the three quarks as blue, green, and red. These three colours combined result in the colour white. Consequently, the allowed arrangement of the triplicity is such that a string of three quarks has to be white. It is always white for it has to be made of red, green and blue. This indicates that there is no preferential arrangement between the three quarks within the most inner layer of matter. It may be said that there is no inner hierarchy. The inner structure of matter is fixed. Matter may only be assembled into more and more complex structures, but it cannot be transmuted into a finer macrocosmic substance. The first layer within the physical world that provides the possibility for a preferential re-arrangement is the rational mind. The rational mind includes the primordial triplicity in the form of the three ordinary faculties of intellect, heart and self. This is the first and only layer within the physical world that is capable of transmutation.

***

Now we may return to our contemplating man to explain to him his situation and his options. Man may solve his situation by finding a means by which he could gain access to the higher modes of oscillations that are present within the Macrocosm. It turns out, that these means have been invested in another "species" that, at one time, appeared among ordinary men. These extraordinary beings are referred to as Perfect Men.

The Perfect Man

> When Perfect Men appeared on earth to achieve closer control of evolutionary trends, certain ordinary men were initiated by them: that is, ordinary men were given access to a technique whereby their minds could become able to process conscious energy and hence to achieve contact with the Absolute's intention.
> (*Ernest Scott*)

The Cosmos is arranged according to a universal design that is based upon the principle of hierarchy. This hierarchy is often compared to the patterns observed among plants, animals, and heavenly bodies. For example, the rose to the flowers bears the same relation as the oak to the trees, or the honeybee to the insects, or the eagle to the birds, or the lion to the beasts, or the Sun to other heavenly bodies. The Perfect Men, however, are just as much a separate species among the various kinds of men, as man is a separate species within other forms of organic life. Ordinary man is superior to animals; so too the Perfect Man is superior to other men by virtue of the refinement of his mind.

Long before the actual creation of humanity, the Absolute created an impression of the Perfect Man. The Perfect Man must experience the state of the original triplicity. This impression was endowed with that experience; this experience is the normative cognition of the Perfect Man. The impression of the Perfect Man was conceived within the world of ideas, in close proximity to the

Absolute. This impression was the very first idea that appeared after the Absolute had expressed the wish "to be known." After its appearance in the world of ideas, the impression of the Perfect Man was projected onto the world of symbols. It is this projection onto the world of symbols that is recorded in the second Chapter of *Genesis.*

It is important to notice that *Genesis* describes two different sequences of the creation of man. The first Chapter of *Genesis* describes the sequence of the creation of ordinary man within the physical world, i.e., the sequence which corresponds to that developed by science. This is the first part of the ascending loop of creation. It concerns the creation of the physical world. It starts with the Big Bang and follows with the creation of the stars, the Sun, the Earth, the mineral world, the Moon, the vegetal world, the animal word, and it finishes with the appearance of mankind. As far as the degree of sophistication of the created forms is concerned, it is an ascending sequence; it starts with elementary particles and ends-up with man.

The description given in the second Chapter of *Genesis*, however, refers to an earlier stage of the process. Namely, it describes a prior phase, i.e., the descending part of the process that took place before the creation of the physical world. This is why the sequence of events described in the second Chapter runs in reverse order with respect to that described in the first Chapter of *Genesis.* This reverse sequence applies to the events taking place in the Macrocosm prior to the creation of the Universe. In this sequence, the various living forms appear in the opposite order to that manifested on the Earth. i.e., they start with the most advanced system, i.e., man:

> And the Lord God formed man of the dust of the ground, and breathed into his nostrils the breath of life; and man became a living soul.
> (*Genesis 2: 7*)

When the "living soul" appeared, there were neither plants nor herbs:

> And no plant of the field was yet in the earth, and no herb of the field had yet sprung up.
> (*Genesis 2: 5*)

This means that the events described in the second Chapter of *Genesis* are taking place entirely in the world of symbols; they happened before the Universe was created. This sequence applies to the forms that were envisaged within the world of symbols. In this sequence, the human mind ("a living soul") appeared first. The "living soul" refers to the inner faculties of the mind. It is a result of the projection of the divine "dot" onto the world of symbols. At that time, the symbols of plants and herbs were not conceived yet.

The garden in Eden, i.e., a symbolic representation of the world of symbols, came into being later; it was conceived after the formation of man's "living soul."

In the midst of the garden appeared "the tree of life" and "the tree of the knowledge of good and evil," obviously symbolic structures. They belong to the world of symbols. In other words, it was only after the appearance of the human's soul that the garden of the world of symbols was populated with symbol-images of animals and birds:

> The Lord God formed every beast of the field, and every fowl of the air.
> (*Genesis 2: 19*)

The Biblical description of "paradise," with angels and the tree of knowledge, etc., is an allegorical description of the world of symbols.

The man of the second Chapter of *Genesis* is the projection of the Perfect Man onto the world of symbols. As was mentioned earlier, the impression of the Perfect Man was already fully developed within the world of ideas. In the Koran, the impression of the Perfect Man is symbolically referred to as "Mohammed."

While in the world of symbols, Adam was not formed yet as a physical man; he was a symbol of the Perfect Man. Therefore, it may be said that when the idea of the Perfect Man was fully developed within the world of ideas, Adam was "between the water and the clay," i.e., he was not created yet. Here is a saying of Mohammed that alludes to this sequence:

> I was a prophet when Adam was between the water and the clay.

The second Chapter of *Genesis* provides a crucial piece of information about the nature of the Perfect Man. It illustrates the process of Adam's "training" for his future role as a spiritual guide of ordinary men. The training consisted of experiencing the veiling of the inner layers of man's mind. First, Adam experienced the operation of a developed mind. This is symbolically indicated as Adam knowing all the names constituting the world of symbols and the world of ideas, i.e., he was familiar with the overall structure of the Macrocosm. As a projection of the impression of the Perfect Man, he knew it, he comprehended it. His knowledge was superior to that of "angels." Here is Rumi's description of that situation:

> In the body three cubits long which He gave him was displayed everything that was contained in the tablets of destiny and the invisible worlds. He taught Adam the Names. Then, Adam gave instruction to the angels concerning everything that shall come to pass unto everlasting. The angels became beside themselves in amazement at his teaching, and gained from him knowledge other than they possessed before. The revelation that appeared to them from Adam was not contained in the amplitude of their dwellings. In comparison with the spaciousness of the range of Adam's mind, their expanse of comprehension was much narrower. Then the angels were saying to Adam, "Before this time we had a friendship with thee on the dust of the earth. Marvelling what connexion we had with that dust, inasmuch as our nature is of heaven. O Adam, that friendship was owing to the scent of thee, because earth was the woof and warp to create the texture of the fabric of thy body. From this place thy earthly body was woven, in this place thy pure light was found. We were in the earth, and heedless of the earth, heedless of the treasure that lay buried there."
> (*Mathnawi, Book I, 2647-66*)

After experiencing the states of higher consciousness, Adam was put in a state of sleep:

> And the Lord God caused a deep sleep to fall upon the man.
> (*Genesis 2: 21*)

While "asleep," the inner layers of Adam's mind were veiled. Symbolically, this is illustrated as Adam being separated from Eve ("a woman"). Eve represents Adam's inner mind, his inner "jewel." The "jewel" was taken from inside of his body, from the very marrow of his bones:

> And the rib, which the Lord God, had taken from the man, made he a woman.
> (*Genesis 2: 22*)

In this allegorical representation, Eve represents a ray of divine consciousness, the projection of the divine "dot." She provides a link between Pure Essence and Pure Intellect. In the following quote, Rumi indicates that "a woman" is often used by poets as a symbol of a ray of divine consciousness:

> She is a ray of God, she is not that earthly beloved: she is creative, you might say she is not created.
> (*Mathnawi, Book 1, 2437*)

This description of Adam being separated from his inner Eve is an allegorical illustration of the veiling of the human mind from its most sublime part. Afterwards, this "precious jewel" remained hidden. Shakespeare's reference to the veiled mind appears in *As You Like It*, when the exiled Duke praises nature ("adversity") and his exile in it:

> Sweet are the uses of adversity,
> Which like the toad, ugly and venomous,
> Wears yet a precious jewel in his head;
> (*As You Like It, II.1*)

When Adam became separated from his inner Eve, his mind became disintegrated and took the form of the three disconnected faculties of intellect, heart, and ego-self. When Adam woke up, he himself became a symbol of the intellect faculty and Eve became the symbol of the heart faculty. At the same time, his ego-self gained its disturbing prominence. The presence of this disturbing ego was a projection of the flawed states within the world of ideas. In *Genesis*, the ego-self is allegorically presented as a serpent. Just as in the case of ordinary man, the ego-self ("serpent") exercised its influence on the intellect ("Adam") via the faculty of emotion ("Eve"). This is why Adam lost his ability to differentiate between right and wrong. In that ordinary state, Adam was no longer capable of accessing the fruits of "the tree of the knowledge of good and evil."

Separation of Adam from Eve was part of the descending phase of the process of creation. This was a necessary step of his descent onto the physical world. Adam, the Perfect Man, was transformed into the symbol of an ordinary man. It was this symbol of the veiled mind that served as a mould for the appearance of ordinary men. The symbol of ordinary men was not endowed with the experiences of Adam. At the moment of its inception, this symbol consisted of the veiled inner faculties.

Adam of the second Chapter of *Genesis* was the first Perfect Man to appear on the Earth. While in the world of symbols, he tasted the operation of the subtle faculties. Then, his state was reduced to that of ordinary men. In this way he experienced the challenges that ordinary men are faced with. However, his previous experiences left a permanent mark in his mind. This mark was manifested as his inherent predisposition towards the Macrocosm.

Afterwards, Adam, the first Perfect Man, had to appear on the Earth among ordinary humans. This final step is described in *Genesis*, again symbolically, as the expulsion of Adam from the Garden of Eden. Definitively, this was not a sort of punishment. Quite to the contrary, it was a very necessary step of the process. In other words, Adam did not have a choice. Adam had to experience the pain of separation before he appeared in a mortal form among ordinary men. It was his function to be "expelled from Paradise." Before he was sent onto the material world, he experienced and learned how it was possible to bring ordinary men to the completion of their ultimate purpose.

This particular stage of the creation and the function of the Perfect Man as a Guide are referred to in the following description:

> Man is a symbol. So is an object, or a drawing. Penetrate beneath the outward message of the symbol, or you will put yourself to sleep. Within the symbol there is a design which moves. Get to know this design. In order to do this, you need a Guide. But before he can help you, you must be prepared by exercising honesty towards the object of your search. If you seek truth and knowledge, you will gain it. If you seek

> something for yourself alone, you may gain it, and lose all higher possibilities for yourself.[39]

In other words, Adam's adventures in "Paradise" were part of his training. He was being prepared for this function as the first Guide of humankind.

***

What is important to notice is the fact that at the time of Adam's appearance on the Earth, mankind had already multiplied and been physically fully developed. Adam was a different man than those around him. Adam's inner being had already been exposed to the world of symbols; he had experienced the operation of the creative mind. All these experiences and feelings were completely unknown and foreign to people around him. Because he himself experienced them, he was able to understand their situation. His role was to help ordinary men to understand the importance of recognizing the operation of the ego-self and provide them with a recipe on how to reform their rational mind.

Adam of the second Chapter of *Genesis* represents a different species among ordinary men. He was the first in the line of the Perfect Men who were needed to assist and oversee the ascending part of the creative process. His role was to guide ordinary men on their journey towards the Absolute. At this point, the process was passed to man. This is why the Absolute could rest "on the seventh day."

---

[39] *The Way of the Sufi*, Idries Shah, 1980, p. 263 (See Note #16).

Up to this point, man had played only a passive role in the process. Starting from the moment of Adam's appearance, man would have to make deliberate efforts in order to initiate, sustain and continue the evolutionary process. This was the beginning, or the genesis, of the second part of the ascent. This phase of the process is called "deliberate evolution" or "conscious evolution." It marked the end of the mechanical creation. Man was faced with an incredibly difficult and demanding challenge.

Before Adam could appear in his corporeal form, mankind and the entire earthly environment had to be properly prepared. There is a story in Rumi's *Mathnawi* which illustrates how the "observers" from the Macrocosm were surveying the conditions on the Earth. These observers are symbolically referred to as angels. There were four angels who were visiting the Earth. Their arrivals corresponded to the four phases of the formation of the Earth. Namely, the first visit led to the collapse of the wavefunction of the mineral worlds (*constructive mode*); the second visit led to the formation of the vegetal world (*vital mode*); the third visit led to the animal world (*automatic mode*); the fourth visit marked the readiness for the appearance of mankind (*rational mode*). As always in this type of stories, the intended meaning becomes apparent when one is not following what appears to be the main trend of thought:

> When the Almighty determined to create mankind, the angel Gabriel was sent to bring back a handful of earth for the purpose of forming Adam's body. Gabriel accordingly proceeded to the Earth to execute the Divine command. But the Earth, being apprehensive that the man so created would rebel against the Maker and draw down a curse upon her, remonstrated with Gabriel. She asked him to leave her and go away. At last Gabriel granted her wish and returned to heaven without taking a handful of earth. Then God sent the angel

Michael on the same errand. When Michael reached the Earth, he put forth his hand to seize some clay from her. The Earth trembled and started pleading with him, shedding tears and making similar excuses. Michael also listened to her crying and returned to heaven with empty hands. Then God said to the angel Israfil, "Go, fill thy hand with that clay and come back." But Israfil also was diverted from the execution of the request. At last God sent Azrael, the angel of firm resolution and strong mind. Again, the Earth pressed him with arguments to take pity on her request. Azrael, who was of sterner disposition than the others, resolutely ignored the Earth's entreaties. He told her that in executing this order, painful though it might be, he was to be regarded only as a spear in the hand of the Almighty. On the return of Azrael to heaven with a handful of earth, God said he would make him the angel of death. Azrael was worried that this would make him very hateful to men. But God said that Azrael would operate indirectly, i.e., through disease and sickness, and men would not look for the prime cause beyond these secondary effects. Moreover, death is in reality a boon to the wise men. Only fools would demand that there was no such thing as death.
(*Mathnawi, Book V, 1556-1709*)

The visits by the four angels were needed to make sure that the earthly environment was ready for the appearance of mankind. It was only once the conditions were correctly prepared, that mankind could appear there. The interesting thing is that the environment was ready at the time of Azrael's visit. Azrael is the angel of death. This means that from the very beginning, the appearance of man was entangled with his physical death. Just like the creation of the Universe. At the moment of its creation (the Big Bang), the Universe was immediately entangled with its disappearance.

Adam could appear on the Earth only after the complete formation of man's brain. The formation was completed when a physical interface allowing for the activation of the inner faculties was implanted within the brain. This was the final accommodation of the human brain to the highest zone of the field of consciousness available in the physical world; the last stage of the mechanical creation. One may look at this last stage as the final formation of the human DNA. According to a recent suggestion, the last stage of the creation was completed some 40,000 – 60,000 years ago (approx.) This last physical stage was identified as an adaptation of the brain that allowed for a simultaneous operation of the left neocortex and the right neocortex. Although the inner faculties remained in their latent states, there was a certain by-product of that final adjustment. Namely, it triggered man's abstract imagination. This event has been described by Joe Griffin and Ivan Tyrrell in their book *Godhead: The Brain's Big Bang*:

> …This triggered an explosion of creativity that fostered complex languages, abstract thought, ingenious tool-making, the production of 'art object' - carving, drawing, painting, decorative clothing - and spiritual symbolism. …
> It is our view that the template for modern humans could not possibly have developed piecemeal through incremental small advances prior to the Upper Palaeolithic period, as some scientists believe. This is because the appearance of creativity would have involved our ancestors free-associating and imagining things that weren't in front of them. To do this, they would have had to go into the REM state in their right neocortex (what we also call the 'psychotic mind') and actively daydream, and that is dangerous because doing so has the potential to unleash schizophrenia. The only way this could have been done is if, hand in hand with accessing the REM state, they also accessed reason, logic and focused attention,

> which are functions of the left neocortex. Unless both the left and right brains were switched on *simultaneously*, the daydream revolution could not have happened. This is because if one hemisphere gained dominance slowly over thousands of years it would produce either a schizophrenic creature or an autistic one. In either case, humankind would not have survived.[40]

This event marked the beginning of abstract thinking and artistic creativity. This final accommodation within the physical world was driven by a field effect, i.e., it was caused by a change within the field of consciousness. Therefore, it had an impact not only on man but also on the earthly environment. Its side effect has been imprinted in nature and affected the animal and mineral worlds. This explains the artistic inclination of Architect, the bowerbird. The decorative skills of the bowerbird may be just a frozen mark of that event that spilled-over onto the animal word. A natural formation of rocks in la Cote de Granit Rose in Brittany, France, is another mark of that event left on the mineral world. Carved by waves, wind, ice, tides, and salt, the pink granites took on shapes of strange plants, animals, and human-like forms.

The Brain's Big Bang marked the completion of the period of mechanical evolution. In other words, mechanical evolution was realized between the Big Bang and the Brain's Big Bang.[41] Afterwards, man was fully prepared for the next phase of the process: deliberate evolution. Mankind was ready for the appearance of the first Perfect Man.

---

[40] *Godhead: The Brain's Big Bang,* Joe Griffin and Ivan Tyrrell (HG Publishing, Chalvington, United Kingdom, 2011, p. 197).

[41] As indicated in the previous Chapter, deliberate evolution was initiated some 10,000 years ago (approx.). This means that Adam appeared among ordinary men 30,000 – 50,000 years after the Brain's Big Bang.

As indicated previously, on his own, man would not be able to fulfil his mission. Mankind needed access to higher modes of consciousness. First of all, however, he needed some guidance and assistance. Physically, men were in the same state as that of Adam. But Adam had experienced the higher states; he had already travelled within the world of symbols. Therefore, he knew how to regain them.

At this point it is interesting to recall that, according to the Samaritans, all mystical teaching stems from one book: *The Book of Signs*, which Adam supposedly brought with him from Paradise. It was this book that enabled him to have power over the elements and invisible things.[42]

The appearance of Adam marked the beginning of the critical part of the process that was invested in mankind. Since then, the fate of men has been in their own hands.

***

We may think of man's mind as a replica of the Macrocosm. In its natural state, however, there is a veil that separates ordinary "walking" consciousness from the higher states of mind. Ordinary man's consciousness is too raw to perceive the presence of the multiple worlds that surround him. This is why it is said that, in his ordinary state, man is infinitely far away from the Absolute. However, this "separation" does not mean "distance" or "location." There is no other location. Everything is within the human mind:

---

[42] *Oriental Magic*, Idries Shah (The Octagon Press, London, 1992, p. 11).

> We are nearer to him than his jugular vein.
> (*Koran, 50:16*)

Adam was the first in the line of the Perfect Men who, from time to time, have appeared to serve man and save him from his "blindness" by awaking him from sleep. These men are always in touch with the currently projected cosmic matrix. They are like "cosmic doctors" who carry "medicine" for men's mental amnesia. These "cosmic doctors" often live in the world almost unnoticed. They have been of all races, and they have belonged to all faiths.

These "cosmic doctors" are in the genetic line of Adam. They may be thought of as the inheritors of Adam's spiritual DNA. It is in this sense that they are like a different species among ordinary men. Unlike Adam, however, they have to experience voluntarily the "expulsion from the Paradise." In an earlier part of their life, each of them has to follow his natural inclination to complete his personal journey into the Macrocosm. As each one ascends to his specific destination within the Macrocosm, he has to detach himself from earthly indulgencies. His destination within the invisible world is determined by the evolutionary needs of the community among which he is to work after his return. During his ascent, his self-faculty is gradually "dissolved". Afterwards, he has to return onto the Earth. As he is descending, he reintegrates his self with a composition which is different from the original one. This change is the result of the experiences and knowledge that he has gained during his journey. At his return onto the physical world he becomes a new man. He is now transmuted.

While on the Earth, these cosmic doctors are not in their own land, because their own land is beyond the physical world. The changes, which they went through, remain imperceptible to ordinary men.

In other words, they do not belong to ordinary men, and yet they do. Their relationship to others is like that of refined gold compared to ore. This means that although their outward form and even a part of their essence may be visible, their whole depth only unfolds to those who are developed enough to understand and perceive it:

> All the nuances of this "stranger-hood" have to be felt if the paradox of the "saviour from afar" is to be apprehended and made to work within the organism in which he appears.[43]

The process of divine synthesis requires a spectrum of Perfect Men within the Macrocosm. At the top of this spectrum is the unperturbed impression of the Perfect Man. The reality of the Perfect Man is the first cognition of the Absolute as itself; the fulfilment of the purpose of the creation.

At the lower level of this spectrum is the symbol of the Perfect Man that is represented by "Adam." Within the Macrocosm, there are a number of templates of the Perfect Man which represent the various degrees of consciousness between the level of "Adam" and the level of "Mohammed." In scriptures, these various degrees of the Perfect Man in their earthly manifestations are referred to as the great prophets and messengers. These various manifestations of the Perfect Man appeared on the Earth at different historical times. Their appearances were associated with the key milestones of man's voluntary evolution. At the same time, they were making available the various modes of oscillation of the universal field of consciousness. These modes of consciousness were made available to ordinary men at different times in the Earth's planetary history.

---

[43] *The Commanding Self*, Idries Shah (The Octagon Press, London, 1994, p. 35).

Each mode of oscillation was needed for the activation of a specific inner faculty of the human mind; each mode of consciousness was higher in its developmental potential than the one before. Each new mode of oscillation needed a guide who would act as the facilitator of these new capabilities.

The guides' role is to prepare and assist men in the proper assimilation of these new evolutionary potentialities. These gradually increasing evolutionary potentialities are symbolically alluded to in the following words written by Omar Suhrawardi, a 13th century Persian author:

> The seed of Divine Wisdom
> was sown in the time of Adam
> germed in the time of Noah
> budded in the time of Abraham
> became a tree in the time of Moses
> gave fruits in the time of Jesus
> and produced pure wine in the time of Mohammed.[44]

The appearance of Adam, the first Perfect Man on the Earth, was related to the event when man's brain was provided with an interface allowing for the activation of the inner faculties of the mind. At that time, mankind was faced with the incredibly difficult challenge of copping with the sudden explosion of imagination and intellectual powers. This was the time for the taming of the ego-self, the faculty which was further empowered by these new capacities of the human brain. It was the time for the re-arrangement of the inner hierarchy within the ordinary faculties of

---

[44] Quoted in the Editor's Note to *The Authentic Rubaiyyat of Omar Khayaam*; translated by Omar Ali-Shah (IDSI, Los Angeles, CA, 1993, p. 7).

the intellect, the heart, and the self. Adam was perfectly prepared for this task as he had previously experienced himself these challenges in the world of symbols.

The next evolutionary milestone was marked by the appearance of Noah. This was the time for the formation of the creative mind. This milestone required the activation of the Spirit faculty and the Secret faculty. This is symbolically indicated in the story of Noah and his three-level ark, which is a representation of the Cosmos consisting of the animal world, mankind, and the Macrocosm.

The next Perfect Man appeared when it was the time for the formation of the sublime mind. The appearance of Abraham marked the experience of the Mysterious faculty.

The appearance of Moses was the mark of the availability of perceptions associated with the Concealed faculty.

Jesus' appearance marked the first experience of the highest form of the ascending triplicity. In Jesus' tradition this encounter is referred to as "three deities":

> Jesus said, "Where there are three deities, they are divine.
> Where there are two or one, I am with that one."
> (*The Gospel of Thomas, 30*)

In this way Jesus indicated that he himself represented the highest form of triplicity. He knew that his disciples were incapable of recognizing that. He was helping them, therefore, by saying that although they cannot see these "three deities," they have been exposed to all of them through his presence.

The appearance of Mohammed marked the first arrival of a man within close vicinity to the Absolute. As a result, ordinary man was provided with the potential of arriving at his ultimate destination.

Each of these six spiritual millennia is traditionally associated with an individual, e.g., Adam, Noah, Abraham, Moses, Jesus, or Mohammed. This does not mean that in each of those epochs there was only one individual, who was exposed to that specific evolutionary experience. In reality, each of these individuals is a representative of a group of people who went through the same or a similar experience at that time. Jesus' comment recorded in *The Gospel of Thomas* alludes to this fact:

> Jesus said, "From Adam to John the Baptizer, among those born of women, there is no one greater than John the Baptizer./…/ But I have said that whoever among you becomes a child will know the kingdom and will become greater than John."
> (*The Gospel of Thomas, 46*)

These were the six spiritual millennia of the ancient world during which a series of evolutionary potentialities was made available to man. Since then, it has been up to man to make full use of his potential. This is why it may be said that man originates from far away; so far that in speaking of his origin, one may use such phrases as "beyond the stars." Some of his feelings and attractions to natural forms of beauty are indicators of this origin. In mystical literature these various forms of attractions are often referred to as "love." In his natural ordinary state, man is attracted to forms that belong to the lowest zone of the field of universal consciousness. These forms are manifested in the physical world as stars, flowers, birds, landscapes, other people. His attractions are driven by his physical senses of sight, touch, hearing, taste, and smell. Other types of attraction are stimulated by his faculties of self (greed, dominance), heart (sensuality, emotionality), and intellect (science, art). All these attractions are limited to things which are manifested within the physical world. As man develops his inner faculties, his attractions expand to the various forms and qualities contained within the invisible worlds. In this way his repertoire of "love" becomes further enriched. This phase of the process may be compared to rarefications of the true reality that is being projected from the Realm. There is a sequence of successive rarefications, each appearing as absolute in its own field of perceptivity of beauty.

***

Mohammed's experiences during the Night Journey provide important details that greatly help to understand the dynamics of the entire Cosmos. According to this tradition briefly mentioned in the Koran, he made a journey across the entire Macrocosm. During his ascent, he was shown a ladder, which he climbed with the angel Gabriel through the seven stages of heaven. On his way he met several prophets, including Abraham, Moses, and Jesus. Gabriel was not allowed to pass the seventh heaven. Afterwards, Mohammed travelled alone. At one point he ended up being in close vicinity to the Absolute. This close vicinity is described symbolically in the Koran as "two bow-lengths," i.e., within a circle formed from two bows:

> At a distance of but two bow-lengths or (even) nearer
> (*Koran, 53:9*)

The distance of "two bow-lengths" indicates that the human experience, which is symbolically represented by "Mohammed," was not fully completed yet at that time. Mohammed was not annihilated in the presence of the Absolute. There were still "two bow-lengths" of earthly traces attached to him. In order to completely purify his mind, Mohammed, like every prophet before him, had to return onto the Earth and complete his service to mankind. Only after his physical death, was he able to continue his journey up to his ultimate experience.

After Mohammed's death, the function of the custodians of the human evolution was vested into a line of Guides who are needed to sustain the development of mankind. Their role is that of facilitators of the activation of the inner faculties of the human mind. The Guides, as the prophets before them, have to experience the arc of ascent (their personal "night journey") and then return to discharge their service. As a result of these experiences, they gain access to a set of modes of the field of consciousness. The level of their ascent is correlated with the needs and potentialities of the communities among which they live. They act as active *observers* for their community and their function is to direct the evolutionary process. The Guides are capable of collapsing nodal zones of higher modes of the field of universal consciousness. The collapsing zones may be focussed on an individual, a group of people, an event, or it may be temporarily attached to a specially designed object or structure. Such a capacity of the Guide is referred to as giving "blessing" or invoking "baraka." In the case of an individual, a Guide may activate his inner faculties in accordance with that person's potentiality and only when this particular person has correctly prepared himself for such an experience. However, the preparation does not require the complete subduing of the ego-self. A Guide may bring certain individuals over the buffer zone. This experience may be induced prior to these persons' physical death. In this way, these persons' spiritual journey may be greatly accelerated.

By providing access to higher oscillations of the field of universal consciousness, a Guide helps to pass through the barrier between the physical world and the invisible worlds. This is possible, because the Guide is present in the physical world but his mind operates within the Macrocosm, i.e., on the other side of the buffer zone. He has access to the modes of the higher oscillations of the universal field of consciousness. Therefore, he is able to activate the oscillations that correspond to the inner faculties. In this way, a person's mind may be elevated to a higher state.

There is an analogous effect that was discovered by modern physics. It is known as the quantum jump or the atomic electron transition. In this effect, an electron can jump from a lower orbit onto a higher orbit by absorbing an external impulse of energy (photon). In this way, the energy of that electron is discretely increased.

Inducing the activation of the inner faculties of the mind may be compared to grafting trees or flowers. A Guide, therefore, is able to ennoble another person by engrafting a sample of evolutionary matrix onto his inner being. This is why a Guide may be compared to a gardener who, by his skills of grafting, is able to grow new species in his garden. Such engrafting allows one to gain immortality, therefore it is often referred to as overpowering of "time." Shakespeare refers to this process in his sonnets. In Sonnet XV, the poet's mentor[45] indicates that he may help him to win the "war with Time":

> And all in war with Time for love of you
> As he takes from you, I engraft you new.
> (*Sonnet XV, 13-14*)

---

[45] There are two voices in the Sonnets, i.e., the poet and his mentor - see *Shakespeare's Sonnets or How heavy do I journey on the way*, W. Jamroz (Troubadour Publications, Montreal, 2014, p. 7).

An individual, however, has to be able to recognize his Guide in order for the Guide to induce constructive experiences. Recognizing a Perfect Man is possible only when the postulant, man or woman, is "sincere." To reach this stage of sincerity, man has to learn to set aside automatic assumptions based upon rules employed in testing different types of phenomena. Here is an example of such recognition. It was recorded in an episode from Jalaluddin Rumi's life:

> One cold November morning, the mysterious dervish Shams was standing in the front of the Inn of the Sugar Merchants in Konya. At that moment, Jalaluddin Rumi happened to be passing by. Rumi sat on his horse as his students scrambled to walk beside him and hold his stirrup. He had just finished giving a class at the College of the Cotton Merchants.
> Shams jumped from the crowd, seized the bridle of the horse and shouted, "Tell me, was Mohammed the greater servant of God, or was it Bayazid of Bistam?"
> Rumi felt the eyes of Shams look past his own into the very essence of his being, causing streams of energy to flow within his body. Rumi replied, "Mohammed was incomparably the greater – the greatest of all Prophets and Saints."
> Then Shams said, "How is it that Mohammed said, 'We have not known Thee as Thou rightly should be known,' whereas Bayazid said, 'Glory onto me! How very great is my Glory'."
> Hearing this, Rumi fainted. When he recovered, he explained: "Bayazid's thirst was quenched by only one cup, and his capacity was satisfied by one draught; whereas the thirst and capacity of the Prophet was limitless and was beyond measure."[46]

---

[46] This version of the story was extracted by the author from *The Whirling Dervishes*, Shems Friedlander (State University of New York Press, Albany, NY, 1992, p. 45).

The point of this encounter was that when Sham asked the question, Rumi did not know the answer. Sham was able to expose Rumi to a higher level of consciousness. When Rumi returned from this brief "journey," he knew the answer. It was in this manner that Sham chose Rumi as his pupil and Rumi recognized Sham as his guide.

The story "The Sultan Who Became an Exile" gives an example of how a Guide is able to induce an extraordinary experience in a man. The quality of such an experience, however, is a function of that man's readiness. As it is explained in the conclusion of the story, in the case of an unprepared man, such an experience is of limited value:

> A SULTAN of Egypt, it is related, called a conference of learned men, and very soon -as is usually the case- a dispute arose. The subject was the Night Journey of the Prophet Mohammed. It is said that on that occasion the Prophet was taken from his bed up into the celestial spheres. During this period he saw paradise and hell, conferred with God ninety thousand times, had many other experiences - and was returned to his room while his bed was still warm. A pot of water which had been overturned by the flight and spilled was still not empty when the Prophet returned.
>
> Some held that this was possible, by a different measurement of time. The Sultan claimed that it was impossible.
>
> The sages said that all things were possible to divine power. This did not satisfy the king.
>
> The news of this conflict came at length to the Sufi sheikh Shahabudin, who immediately presented himself at Court. The Sultan showed due humility to the teacher, who said: 'I intend to proceed without further delay to my demonstration: for know now that both the interpretations of the problem are

incorrect, and that there are demonstrable factors which can account for traditions without the need to resort to crude speculation or insipid and uninformed "logicality".' …

Now the sheikh ordered a vessel of water to be brought, and the Sultan to put his head into it for a moment. As soon as he had done so, the Sultan found himself alone on a deserted seashore, a place which he did not know.

At this magic spell of the treacherous sheikh he was transported with fury, and vowed vengeance.

Soon he met some woodcutters who asked him who he was. Unable to explain his true state, he told them that he was shipwrecked. They gave him some clothes, and he walked to a town where a blacksmith, seeing him aimlessly wandering, asked him who he was. 'A shipwrecked merchant, dependent upon the charity of woodcutters, now with no resources,' answered the Sultan.

The man then told him about a custom of that country. All newcomers could ask the first woman who left the bath-house to marry him, and she would be obliged to do so. He went to the bath, and saw a beautiful maiden leaving. He asked her if she was married already. … She said that she was not married, but pushed past him, affronted by his miserable appearance and dress.

Suddenly a man stood before him and said: 'I have been sent to find a bedraggled man here. Please follow me.'

The Sultan followed the servant, and was shown into a wonderful house in one of the sumptuous apartments he sat for hours. Finally four beautiful and gorgeously attired women came in, preceding a fifth, even more beautiful. She, the Sultan recognized as the woman whom he had approached at the bath-house.

She welcomed him and explained that she had hurried home to prepare for his coming, and that her hauteur was only one of the customs of the country, practised by all women in the street. …

The Sultan stayed seven years with his new wife: until they had squandered all her patrimony. Then the woman told him that he must now provide for her and their seven sons.

Recalling his first friend in the city, the Sultan returned to the blacksmith for counsel. Since the Sultan had no trade or training, he was advised to go to the marketplace and offer his services as a porter.

In one day he earned, through carrying a terrible load, only one-tenth of the money which was needed for the food of the family.

The following day the Sultan made his way to the seashore again, where he found the very spot from which he had emerged seven long years before. Deciding to say prayers, he started to wash in the water: when he suddenly and dramatically found himself back at the palace, with the vessel of water, the sheikh and his courtiers.

'Seven years of exile, evil man!' roared the Sultan. 'Seven years, a family and having to be a porter! Have you no fear of God, the Almighty for this deed?'

'But it is only a moment,' said the Sufi master, 'since you put your head into this water.'

The courtiers bore out this statement.

The Sultan could not possibly bring himself to believe a word of this. He started to give order for the beheading of the sheikh. Perceiving by inner sense that this was to happen, the sheikh exercised the capacity called *Ilm el-Ghaibat*: The Science of Absence. This caused him to be instantly and corporeally transported to Damascus, many days' distance away.

From there he wrote a letter to the king:

'Seven years passed for you, as you will now have discovered, during an instant of your head in the water. This happens through the exercise of certain faculties, and carries no special significance except that it is illustrative of what can happen. …

'It is not whether a thing has happened or not which the important element is. It is possible for anything to happen.

> What is, however, important, is the significance of the happening. In your case, there was no significance. In the case of the Prophet, there was significance in the happening.'[47]

The Sultan's adventures demonstrate his current inner state. In this state, the Sultan is not able to survive within the invisible world ("a place which he did not know"). Although there are available modes of higher consciousness (women who are "not married"), he is not capable of developing his inner faculties. This is symbolically illustrated as his inability to provide for his wife and his "seven sons."

***

The Perfect Man has to discharge two roles. The first is to organize man in a safe, just and peaceful manner, to establish and help sustain communities. A correctly functioning community allows and stimulates ordinary men to adjust their actions in such a way as to harmonize themselves with that community. This provides the external template of a balanced mind and helps in the formation of a correctly operating rational mind. In other words, this is the very minimum that is required for the preservation of the human race.

The second role is inward, to lead people from outward stabilization to the performance which awakens them and helps to make them permanent.

---

[47] *Tales of the Dervishes*, Idries Shah, p. 35 (see Note #27).

The Descent of the Soul

> Ordinary human love is capable of raising man to the experience of real love.
> (*Hakim Jami*)

It is the state of the human mind that drives history and the welfare of societies. At different historical times, the various nodes of the field of consciousness were purposely activated within select communities in designated geographical areas. The birth and death of each civilization were planned according to the overall cosmic plan. In this manner the various civilizations were developed in succession. The various nodes of the field of consciousness, in their wave forms, were stored by being attached to specially designed architectural structures such as certain buildings, city squares, castles, cathedrals, monasteries, gardens. These structures then served as temporary storage-houses for the specific wavefunctions of the evolutionary spectrum in those designated geographical areas.

Sometimes these structures had unusual shapes that raised all sorts of questions as to the very reason for their existence. For example, Castel del Monte is one of the most spectacular structures in southern Italy. One may look at it as a simplified version of the scientific operator of the "bower" equation. Built from local stone, the massive octagonal crown-shaped footprint has two floors with 16 trapezoid rooms, eight on each floor. There is an octagonal tower on each of the eight corners. Castel del Monte has been described by historians as an "ideal and useless construction" and a "bizarre labyrinth." A similar challenge for historians is presented by the octagonal building known as Tour Evraud in Fontevraud Abbey in France. This building is topped with an octagonal hood and is surrounded by eight round niches. Again, the purpose of this strange structure is still subject to controversy. Some believe that the niches served as gigantic fireplaces and, therefore, Tour Evraud was used as a kitchen; others argue it was a smokehouse.

Such "sensitized" places often became pilgrimage sites because they attracted people. People sensed that there was something special there. By visiting them, people may experience something positive. However, each of these places has a limited capacity to absorb the negativity associated with emotionality and fanaticism that are often brought by huge crowds of visitors. After some time, therefore, such a place may "shut-off" its evolutionary charge. At that point, these places are usually destroyed or turned into venues for arts, entertainment or other commercial activities.

***

To people all over the world, the Taj Mahal, mausoleum of the Mughal Empress Mumtaz Mahal, is synonymous with India. The Taj Mahal, one of the seven wonders of the modern world, is undoubtedly one of the most spectacular buildings. It contains elements of the bower of the bird-of-paradise and some features of the crystal-like geometrical figure of theoretical physicists. But there is something else, something that cannot be reproduced either by nature or the esthetics of mathematical equations.

One of the most striking and fascinating things about the Taj Mahal is that its design is a representation of a soul's journey through the various cosmic strata. The Taj Mahal does not only illustrate this journey; it also allows one to experience it.

In February 2004, I was with friends in Agra, India, where the Taj Mahal is located. Here is my account of such an experience:

> We left our hotel early in the morning, before dawn. We wanted to see the Taj Mahal at sunrise. There already were many people. It was a mixed crowd of tourists, beggars, ancient and decrepit men, mothers with small children asking for money, young boys aggressively trying to sell their wares. It was quite chaotic, annoying and sometimes even irritating.
>
> The entry into the Taj Mahal leads through the arched gate of the red tower. The tower is decorated with calligraphic inscriptions that include the text of the Koranic verse "Daybreak." Metaphorically, the entrance represents the birth of a new soul. Passing through the gateway was like experiencing the transition point between nothingness and arrival within the Macrocosm. Indeed, after passing through the gateway we entered into quite a different world. We saw the garden and then in the distance some outlines of the walls, cupolas and minarets of the Taj Mahal.
>
> The creation of the Taj Mahal dates back to the time of the conquest of India by the Mughal armies. Shah Jahan (1592-1666) was a soldier and a statesman. He was the grandson of the great Akbar (1542-1605).
>
> Shah Jahan spent most of his life on military campaigns. His wife, the beautiful Mumtaz Mahal (the Treasure of the Palace) was his constant companion. In 1631, during one of these military expeditions, Mumtaz Mahal died.
>
> There is an interesting legend regarding the plan of the garden and the Taj Mahal. According to this legend, Mumtaz Mahal, while on her death bed, asked Shah Jahan to build a beautiful

palace-like tomb with a lush garden which she had seen in a dream the previous night. After her death the king sent for the architects. They submitted various plans of the proposed tomb but he did not approve any one of them. There was a wise man in Agra. He brought to the king a unique plan. He presented it to the king and said: "This is the design of the same palace and garden which the queen had seen in her dream. I give it to you to execute her will." The king approved the plan and the Taj Mahal was consequently built. The construction of the Taj Mahal began in 1631 and was completed in twenty two years. A total of twenty thousand people were deployed to work on it.

Recently a new piece of evidence has been discovered that adds a new dimension to the above quoted legend. Namely, it has been found that the design of the garden and the Taj Mahal is based on a diagram that was described by Ibn Al-Arabi, a 12th century Andalusian mystic. In Ibn Al-Arabi's work entitled *Meccan Revelations*, this diagram was related to the Garden of Paradise that he experienced during a spiritual journey. It was this diagram that was the blueprint for what is widely considered to be the most beautiful building in the world.

I stayed in amazement looking at the Taj Mahal. The Taj Mahal is a very large structure. The mausoleum, instead of occupying the central point -as it is in other places in Central Asia- stands majestically at the north end, just above the Jumna river. At the center of the garden, halfway between the mausoleum and the gateway, there is a raised water pool. The pool has been arranged to perfectly reflect the Taj Mahal in its waters. The reflection is so perfect, that it looks like the original, and that the Taj Mahal is its reflection in another world. This pool feeds four water canals that run to the north, south, west and east. They represent the four inner faculties joining together to form the fifth one; the fifth faculty allows one to see Reality. The earthly "body" (the Taj Mahal) is just a partial reflection of that

reality.
Fountains and rows of cypress trees adorn only those water canals that run along the north-south direction. Thus, the visitor's attention is diverted from the sides and is attracted toward the north, where the Taj Mahal resides. In the changing light of the rising sun I could not distinguish any details of the mausoleum building. The garden was filled with the scent of flowers. There were plenty of happily chirping birds. I could feel on my face the cool and wet morning breeze. The Taj Mahal appeared luminous. The white cupolas and white minarets seemed to radiate a light of their own. I stayed there for a while and looked at the Taj Mahal, trying to seize and impress on my memory all the details of the building and of everything else around me. This was a wonderful scene! I had the impression that the Taj Mahal was alive. This entire place was like a living creature. It was breathing, it was moving, it was singing, it was growing; it was changing its colors. At first, just before sunrise, the Taj Mahal was white, barely visible against the blue sky. Then it became reddish in the rays of the rising sun. Then it became yellow. Like a flower it was growing and swinging in the morning breeze. I could have stayed there forever – spellbound by the magnificent view. But it was not just the view. It was the feeling that this image was stirring deeply inside. It seems that the Taj Mahal was designed in such a way, that the arched gateway would force the visitors to pass through a point, where they would find themselves in the focus of all the impacts that radiate there, i.e. images, colors, sounds, scents, breeze, movements. And all these would trigger something deeply inside their beings, something very intimate, harmonious and joyful.
We started to walk towards the Taj Mahal. The Taj Mahal was like a huge magnet that was pulling us toward itself. It was an enthralling journey, like traveling on a zooming device across space, time and impressions. The building of the mausoleum was rising before our eyes; it was getting bigger and bigger. At

the same time the image of the garden was diminishing. Instead, the details of the décor and the design of the building were phased into our view. I could still feel the initial impact that I encountered at the arched gate, but the source of the impact was no longer there; somehow it was gradually veiled. We approached the marble platform on which stands the Taj Mahal with its four minarets at the corners. The portals of the Taj Mahal are adorned with the text of the verse "Ya-Sin," with its powerful words: "Be, and it is." Shah Jahan's calligraphers had performed an amazing optical trick while inlaying this black stone calligraphy in the white marble. It was the same trick of false perspective used by Architect, the bowerbird. The letters have been inscribed densely at the bottom, with little plain surface in between. As the inscription rises, it becomes more and more sparse with more plain surface in between the letters. This makes the lettering appear to be of the same size, at the top and at the bottom.

The outside walls of the Taj Mahal are made from white marble panels. The panels are decorated with carved uncolored flowers. I remember that I was surprised while looking at them. Somehow these elaborate marble panels were quite plain and they even seemed to be not quite completed. They were missing something.

We walked up and came to the doors leading into the interior of the mausoleum. Inside, two stories of eight rooms surround a central chamber. The central chamber is octagonal. The interior of the mausoleum is like a multidimensional geometrical structure. The whole of the inside walls is covered with stone flowers, carved doors, and carved windows. These beautiful flowers were made out of colored stones that were inlaid in white marble. In the middle, surrounded by a carved marble trellis, were two white tombs, in the center the tomb of the Empress, and beside it that of Shah Jahan. The tombs, aligned along the North-South axis, and above them an olive-oil lamp burned in a brass lantern.

The longer I looked, the more clearly I felt that there was something strange there. I had the feeling that there was a message that the Taj Mahal designers had striven to pass to those who would visit this amazing place.
I looked at the olive-oil lamp that burned above the tomb. The light-glimmering above the tomb was small and insignificant in comparison with the luminosity of the Taj Mahal. But this lamp was the only thing alive inside the chamber! It was the only thing that resembled the living creature that I encountered in the garden. All the rest inside the chamber -despite its amazing decoration- was still, frozen. There was no movement, no breeze, and no changing colors. However, there were traces and indications of the outside life. There were the fresh flowers and the dimmed sun light coming from outside through the trellis carved in the side galleries. I had the impression that the chamber, nicely decorated but somehow inanimate, was like the human body. I felt like a soul that has been trapped within a mortal body. Just like this tiny lamp. Although this tiny burning light seemed to be so insignificant and could be so easily extinguished, it was this light that belonged to and was a part of the eternal entity that resided outside. This light was that shining "dot" hidden within man's inner being; this was the true "Treasure of the Palace."
At this point, one of the Taj Mahal guides raised his head and cried in a loud voice: "Allah!"
The effect of the guide's call was just as that described by Peter Ouspensky, who visited the Taj Mahal in 1914. Ouspensky described his experiences in *A New Model of the Universe*:

> His voice filled the whole of the enormous space of the dome above our heads and as it began slowly, slowly to die away, suddenly a clear and powerful echo resounded in the side cupolas from all four sides simultaneously:
>
> "Allah!"
>
> The arches of the galleries immediately responded, but

> not all at once; one after the other voices rose from every side as though calling to one another.
>
> "Allah! Allah!"
>
> And then, like the chorus of a thousand voices or like an organ, the great dome itself resounded, drowning everything in its solemn, deep bass:
>
> "Allah!"
>
> Then again, but more quietly, the side-galleries and cupolas answered, and then the great dome, less loudly, resounded once more, and the faint, almost whispering tones of the inner arches re-echoed its voice.
>
> The echo fell into silence. But even in the silence it seemed as if a far, far-away note went on sounding.[48]

The chamber seemed to be awakened! The guide knew how to bring this place to life. By invoking the name of God he connected this marble shell with the living entity. Within this sleeping body, he activated the link to the garden, a symbolic illustration of collapsing of various modes of consciousness. This earthly chamber was connected with the living entity via the beautiful trellis carved in the side galleries, via red flowers that were brought from the garden, via acoustic channels that provided echoes responding to the guide's cry.
I thought at that moment that I understood the story attributed to Hasan of Basra about a child with a light:

> I asked a child, walking with a candle,
> "From where comes that light?"
> Instantly he blew it out. "Tell me where
> it is gone – then I will tell you where it
> came from."[49]

---

[48] *A New Model of the Universe*, P.D. Ouspensky (Vintage Books, New York, 1971, p. 334).
[49] *The Sufis*, Idries Shah, p. 235 (see Note #13).

I thought that I knew where the light came from and where it went. At this very moment I knew that the source of the light was outside; it was in the garden, with these avenues, cypress trees, fountains. Hakim Sanai's poem provided me with an additional clue:

> When life finally walks out of the door,
> your tattered soul is straight away renewed;
> your form is freed from the bonds of nature,
> and your soul gives back the spirit's loan.[50]

Slowly I started to understand the message of the Taj Mahal designers. It is like a guidebook that they inscribed in this place. It is a guidebook describing the journey of the human soul through the various strata of the Macrocosm. Visiting the Taj Mahal is like following the passage of the human soul. Entering the arched gateway - this is "The Daybreak," it is like experiencing the creation of the soul - "Be, and it is." The passage through the garden - this is the soul's descent to the earthly domain. On its way to the earthly domain it passes through the different levels of the invisible worlds. While passing by the water pool, the soul acquires the divine "dot." Entering the door to the mausoleum – this is the soul entering a human form. Being inside the chamber – this is our earthly life. A life that is so attractive and so beautifully decorated, but these decorations are there to cover the inanimate marble shell and to remind us about its true nature. Then the Taj Mahal tells us that although this life may tatter the soul, the soul is always connected with its divine origin. At one point, the soul is freed from the bonds of nature and, as it returns to its origin, it "gives back the spirit's loan."

We left the chamber and we walked again in the garden. We could see groups of visitors, like a flow of souls that slowly moved along the cypress avenues towards their earthly

---

[50] *The Walled Garden of Truth*, Hakim Sanai, p. 49 (see Note #9).

destination. This was a continuous flow of souls in colored turbans and robes of various colors, yellow, red, white, black and green. This flow of visitors was also a part of the design of the builders of the Taj Mahal, part of their mystical teaching about the link between humanity and eternity.

We walked outside of the complex. We were back in the world that we had left just a few hours before; back to the noisy and chaotic reality. It seemed to me, however, that now this world was somewhat different. Through experiencing the Taj Mahal, we were somehow enriched. We all carried with us an invisible thread linking us to this beautiful and majestic garden. We were taking with us this thread, in whatever form, to the streets of Agra, to our homes, families, and friends.

Now I knew that the Taj Mahal was a place to experience, to feel and to learn from.

It is through instruments and devices like the Taj Mahal that man is exposed to impacts that can make him aware of his evolutionary potential. If man manages to recognize his potential, he will be able to increase his existence infinitely. If he does not, he might dwindle to a vanishing-point.

Destiny

> You must prepare yourself for the transition in which there will be none of the things to which you have accustomed yourself.
> (*Al-Ghazali*)

There are two possibilities offered by the current scientific model describing the fate of the Universe. The Universe either will be continuously expanding or the Universe's expansion will be reversed leading to its collapse.

According to the model, it is the density of matter that determines the fate of the Universe. It is this physical property of the Universe that determines its future. The density determines the strength of the gravitational force. The first option assumes the density of the Universe to be less than a certain value that has been defined as the critical density. In this case the gravitational pull will not be enough to stop the Universe's expansion. As a result, the Universe will be expanding so rapidly that the gravitational force can never stop it.

In the second option, the Universe is expanding slowly enough that the gravitational force between galaxies may cause its expansion to slow down. This would eventually lead to a stop of the expansion and reverse the process. In the reverse process, the Universe would start to contract. All matter would begin to travel inwards, accelerating as time passes. At the end, all matter would collapse into black holes, which would then coalesce, producing a big crunch. This option requires a spatially finite space and time. Recent experimental evidence suggests, however, that the expansion of the Universe is not being slowed down; it is rather accelerating.

However, there is a significant problem with the estimation of the density of the Universe. If one adds up the masses of all the stars in all galaxies, the total value is less than needed to explain even the currently observed dynamics of the Universe. The calculations indicate that many galaxies should fly apart instead of rotating, or should not move as they do, or should not be formed at all. It has been concluded, therefore, that there is some form of an unknown *substance* that contributes to the overall behaviour of the Universe. In addition to this substance, another type of unknown *energy* is also needed to explain the rate of expansion of the Universe.

Theoretical physicists temporarily solved this problem by assigning the unknown *substance* to the existence of "dark matter" and the unknown *energy* to "dark energy." It is assumed that dark matter and dark energy are abundant in the Universe and have had a strong influence on its structure and evolution.

Dark matter is called "dark" because it does not appear to interact with observable electromagnetic radiation, such as light, and is thus invisible to the entire electromagnetic spectrum. This makes it impossible to detect by using available astronomical equipment.

Dark matter is thought to be composed of some as-yet-undiscovered sub-atomic particles. This means that, as long as the nature and structure of dark matter remain unknown, it is impossible for science to draw a conclusion about the future of the Universe.

***

The model of the Cosmic Oscillator described in this book provides a much clearer picture of the structure of the Universe. According to this model, consciousness is a form of energy. The entire Universe is filled in with the field of universal consciousness. The field of universal consciousness consists of a multiplicity of standing waves. They penetrate every single point of the entire Universe. The nodes of the standing waves are manifested as various forms of matter; the antinodes remain in their wave-like form.

According to Einstein's equation, there is a relationship between energy and matter. Therefore, the model of the Cosmic Oscillator implies that there is a relationship between consciousness (C) and matter (M). This relation may be expressed in the form of a complex number as the sum of two components:[51]

$$C = \alpha M + i \beta M$$

The first component (α M) indicates that part of the universal consciousness that is manifested as physical matter (α – is a normalizing coefficient). The second component ($i \beta$ M) represents that part of the field of consciousness within the Universe which remains in its wave-like form. It may be said that this second component represents "virtual matter" (where "*i*" is the imaginary unit; β is a normalizing coefficient). Virtual matter is a field

[51] Complex numbers are used in physics as a calculation tool. A complex number consists of two parts, "real" and "imaginary." The imaginary part is marked by the imaginary unit "*i*." In quantum mechanics, complex numbers are used to describe wavelike phenomena for particles such as electrons and neutrons.

phenomenon. It is the presence of this virtual matter that has been identified by the physicists as "dark matter" and "dark energy."[52]

According to the physicists' estimate, at the present time, roughly 68% of the Universe is dark energy; dark matter makes up about 27%. This means that ordinary matter makes up only 5% of the entire Universe.[53] Estimating the ratio of ordinary matter and virtual matter at several stages of the formation of the Universe would allow to determine the values of the normalizing coefficients α and β.

The above relationship between consciousness and matter indicates the existence of virtual elementary particles. Just like *relatons*, these virtual particles would act as the intermediaries between matter and the field of universal consciousness.[54] These to-be-discovered particles will be qualitatively much more sophisticated than those discovered so far.

There is another aspect of virtual matter that may be of interest to the cosmologists for it may help to solve another great mystery of the Universe. Namely, the conversion of the lowest grade of consciousness into matter did not stop with the Big Bang. It is a continuous process, although it continues at a much slower rate. This means that the virtual matter is slowly but continuously being converted into ordinary matter. This effect has been observed by cosmologists as a flow of particles known as "cosmic rays." The cosmic rays carry so much energy that cosmologists are still baffled by what object in the universe could have created them.

---

[52] The link between the field of consciousness and dark matter was first suggested by Joe Griffin and Ivan Tyrrell; they introduced the term "subjective matter" to describe dark matter. (*Godhead: The Brain's Big Bang*, Joe Griffin and Ivan Tyrrell, p. 143 – see Note #40).

[53] "Dark Energy, Dark Matter," posted on NASA's web page: https://science.nasa.gov/astrophysics/focus-areas/what-is-dark-energy (November 6, 2019).

[54] The concept of *relatons* is described by Joe Griffin and Ivan Tyrrell in *Godhead: The Brain's Big Bang*, p. 140 (see Note #52).

Consequently, they named the highest-energy particle of the cosmic rays "Oh-My-God particle."

***

According to the mystics, the Universe will not follow any of the options proposed so-far by the physicists. Let's recall that, according to the mystics, the Universe was created to provide an adequate environment for the appearance of man, the observer. This is why corporeality was imposed on him. The corporality was implemented by placing man in the material world. Matter imposes two limitations, space and time. Space was needed to determine a spatial enclosure within which it would be possible to create the conditions needed for the appearance of life. Time was needed to determine the period within which man's task was to be accomplished. This means that the Universe must be finite, both spatially and temporary. The temporal finity of the Universe implies a cosmic "end of time."

The Universe's lifetime is determined within a certain margin to allow for man's errors. In other words, there must be a sort of timing mechanism capable of controlling the Universe's lifetime. This may sound more like a fantasy rather than an even remotely viable possibility. Yet, the existence of such a timing device is conceivable within the presented model of the Cosmic Oscillator. Namely, it is the field of consciousness that acts as the device controlling the lifetime of the Universe. At one point, the process of creation will be reversed: the "end of time" will be initiated by the gradual turning off of the oscillations of the field of consciousness.

At first, the oscillations associated with the highest form of consciousness within the Universe will be terminated: the "dot" will be removed from the human mind. This will mark the end of the human race. Afterwards, there will be no more humans; the human race in its present earthly form will cease to exist. By that future time, the human race will have been transferred into the buffer zone, i.e., the zone that separates the Macrocosm from the physical world. In other words, humanity will be transformed, or "resurrected," into its new form within the buffer zone. In religious texts, this event is referred to as the "Last Judgement."

Afterwards, the oscillations associated with the formation of other organic systems will be gradually turned-off. It is in this manner that the process of the creation of the Universe will be reversed. At one point, the Universe will dissolve to physical "nothingness." That event is alluded to by Shakespeare in Prospero's speech in *The Tempest*, when he dissolves a vision which he has created as an entertainment for his daughter and her lover. He compares this vision to the Universe. Here is Prospero's description of the "end of time":

> Our revels now are ended: These our actors,
> (As I foretold you) were all spirits and
> Are melted into air, into thin air,
> And like the baseless fabric of this vision
> The cloud-capp'd towers, the gorgeous palaces,
> The solemn temples, the great globe itself,
> Yea, all which it inherit, shall dissolve,
> And, like this insubstantial pageant faded
> Leave not a rack behind. We are such stuff
> As dreams are made on; and our little life
> Is rounded with a sleep.
> (*The Tempest, IV.1*)

This situation of the end time is allegorically described in a story entitled "When the Waters Were Changed." Consciousness is sometimes compared to the air we breathe or the water we drink. In this story, the changing of the "waters" corresponds to the switching off of the access to the higher modes of consciousness:

> Once upon a time Khidr, the Teacher of Moses, called upon mankind with a warning. At a certain date, he said, all the water in the world which had not been specially hoarded, would disappear. It would then be renewed, with different water, which would drive men mad.
>
> Only one man listened to the meaning of this advice. He collected water and went to a secure place where he stored it, and waited for the water to change its character.
>
> On the appointed date the streams stopped running, the wells went dry, and the man who listened, seeing this happening, went to his retreat and drank his preserved water.
>
> When he saw, from his security, the waterfalls again beginning to flow, this man descended among the other sons of men. He found that they were thinking and talking in an entirely different way from before; yet they had no memory of what had happened, nor of having been warned. When he tried to talk to them, he realized that they thought that he was mad, and showed hostility or compassion, not understanding.
>
> At first he drank none of the new water, but went back to his concealment, to draw on his supplies, every day. Finally, however, he took the decision to drink the new water because he could not bear the loneliness of living, behaving and thinking in a different way from everyone else. He drank the new water, and became like the rest. Then he forgot all about his own store of special water, and his fellows began to look

> upon him as a madman who had miraculously been restored to sanity.[55]

Ibn Al-Arabi alluded to this particular phase of the process. He mentioned that there will be a time when the last Perfect Man will appear. During his time, sterility will overcome men and women; there will be no bringing forth children. Thereafter men will become as beasts, bereft of feelings and laws. At that future time, the human race will gradually disappear.[56] Symbolically, this description refers to man's evolutionary "sterility," when mankind will become developmentally impotent. Thus, Ibn Al-Arabi indicates that the lowest link of the divine synthesis, of which we are all a part, will come to an end and will be terminated.

***

Let's take a look at how science perceives the future of humanity and what options have been offered by modern physicists.

According to science, the Sun will swell up and engulf the Earth about ten billion years from now. It is within this time frame that an intelligent form of life would have to master space travel in order to escape from the apparent doom.

The overriding idea that has recently been advocated by physicists is to find another habitable planet (exoplanet) on which the human race could continue to exist. The idea of "cosmic" travel as a

---

[55] *Tales of the Dervishes*, Idries Shah, p. 21 (see Note #27).

[56] *The Bezels of Wisdom*, Ibn Al 'Arabi; translated by R.W.J. Austin (Paulist Press, Inc., Mahwah, NJ, 1980, p. 70).

means of preserving the human race has entered the human mind through science fiction. Afterwards, this idea became the driving force of the most recent scientific undertakings. For example, in his speech at the 2017 Starmus Festival in Trondheim (Norway), Professor Stephen Hawking, a leading theoretical physicist of our time, explained this approach in the following way:

> We are running out of space and the only places to go to are other worlds. It is time to explore other solar systems. Spreading out may be the only thing that saves us from ourselves. I am convinced that humans need to leave Earth.[57]

The scientists have realized that, in the last ten thousand years or so, the human DNA has not changed significantly. In other words, we cannot wait for a sort of Darwinian evolution to make us better fitted to handle the upcoming challenges. According to Prof. Hawking, humans should initiate a new phase of what might be called "self-designed evolution." By using genetic engineering, a new generation of DNA will allow prolonging the life to such an extent, that it will be possible to survive lengthy interstellar travels. It is this sort of "engineered evolution" that would allow the human race to survive the unavoidable collapse of our solar system.

Consequently, it is envisaged that there will appear a new race of self-designing automatons. These automatons will be able to redesign themselves in such a way as to meet the upcoming challenges, which they will be capable to predict. These beings will, just as illustrated in numerous science fiction narratives, colonize

---

[57] "Hawking urges Moon landing to 'elevate humanity'," Pallab Ghosh, *BBC News* (June 20th, 2017).

some parts of the Universe. Eventually, these automatons will govern the entire Universe.

Somehow, the concept of "self-designing automatons" sounds similar to Ibn Al-Arabi's developmentally sterile men.

***

There are a couple of questionable assumptions in the above quoted scientific proposal for a prolonged sustenance of humanity. Firstly, it is assumed that the time window for the existence of the human race is defined by the life time of our solar system (e.g., in the range of a few billion years). According to the Cosmic Oscillator model, the lifetime of humanity will be determined by the gradual turning off of the oscillations of the field of consciousness. First, the highest zone of consciousness within the physical world will be switched off; then the lower zones will be gradually terminated. Therefore, the disappearance of the human race will trigger the annihilation of the solar system, and not the other way around. Omar Khayaam, a 11th century Persian poet, alludes to such a hierarchical relationship between humans and the planets in *The Rubaiyyat*:

> Live in no awe of planets. Planets are
> One thousand times more impotent than we.[58]

---

[58] *The Authentic Rubaiyyat of Omar Khayaam (Quatrain #76)*; translated by Omar Ali-Shah, p. 63 - see Note #44 (Omar Ali-Shah's translation is also available in *A Journey with Omar Khayaam*, W. Jamroz, Troubadour Publications, Montreal, 2018).

As indicated in the next Chapter, the human race will start to disappear some ten thousand years from now (assuming that humanity will not eliminate themselves prior to that time by destroying the planet's natural nodal zone of the field of consciousness).

Secondly, the scientists assume that it would be possible to relocate the human race to another star system. This approach ignores the fact that the human race is specific to the nodal zone of the Earth. It is this natural zone of consciousness that provides the needed "environment" for the sustenance of mankind. Although this zone probably extends slightly outside the planet, it does not cover other planets within the solar system. In other words, humans were not made to exist outside of our planet. This means that the presence of humans outside of the Earth would interfere with their natural level of consciousness. This interference is not so much related to the accommodation to different physical conditions such as atmosphere, gravity or cosmic rays. It is rather a side-effect related to being outside of the planet's natural zone of consciousness. An extended presence outside of the planet's zone would have a negative effect on the human mind; it would be destructive. It would be equivalent to the reversing of the process of creation; it would gradually reduce man to an inferior creature. A prolonged presence in space would make men less human, i.e., it would significantly or completely cut them off from the ability to discharge their role of active participation in the evolutionary process. In other words, man's *raison d'etre* would be nullified.

Al-Ghazali, a 11th century Persian philosopher, compares the effect of being cut-off from man's natural zone of consciousness to deprivation of food or certain medicines. The skills of preservation of this vital zone are referred to as "special knowledge":

> The "special knowledge" is that which supports life to such an extent that if its transmission were to be interrupted for three days the kernel of the individual dies, just as someone would die if he were deprived of food, or a patient dies when deprived of certain medicines.[59]

Although this may sound strange and improbable in the context of the various presentations offered by science fiction, such negative effects on astronauts have already been detected. It was reported in a study by NASA that, after spending 340 days aboard the International Space Station, an astronaut's DNA mutated in some of his cells.[60] These genetic mutations affected the astronaut's mental abilities. The effect was measured as a decline in the astronaut's cognitive test scores. So far, the effect of being outside of the planet's natural zone of consciousness has not yet been considered.

---

[59] *Thinkers of the East*, Idries Shah (The Octagon Press, London, 1971, p. 177).
[60] "Scott Kelly Spent a Year in Orbit. His Body Is Not Quite the Same," Carl Zimmer (*The New York Times*, April 11th, 2019).

Death and Rebirth

Had humankind been freed from womb and tomb,
When would your turn have come to live and love?
(*Omar Khayaam*)

Science assumes that man's present structure, mental as well as physiological, is the last word in biological evolution. Furthermore, science regards death only as a biological event. These assumptions have no constructive meaning. These beliefs are in stark contrast to the experiences of the mystics.

According to the experiences of the mystics, the evolution of the human mind is not sealed at the moment of a person's physical death. Quite to the contrary, after death man's journey continues. There are much wider vistas of the Cosmos to be explored. The journey, however, does not lead to another planet, star, or galaxy. Instead, this is a travel through the various layers of cosmic consciousness. The journey starts from that level of consciousness which an individual person has arrived to at the moment of his or her physical death.

Physical death is determined by the turning off of the zone of the field of consciousness that provides a compartment for a person's physical body. The turning off of this zone leads to the disintegration of the body. At this point the *rational mind* is separated from its earthly host. The rational mind starts to drift towards the buffer zone that separates the physical world from the invisible worlds. The duration of travel to the buffer zone will depend on the mind's inner strength. The weak minds will not be able to reach it. Those minds, which are too weak to sustain the link with the "dot," will gradually dissolve to nothingness.

The partially or fully refined rational minds will travel till they eventually reach the buffer zone. The buffer zone is the state which an individual mind is "resurrected" to; it is the intended destination state for most people. It is the buffer zone that science, in its rational and linear thinking, has substituted with an inhabitable "planet" located somewhere in a distant region of the Universe.

At the moment of physical death, an individual's mind experiences a challenging transition. The ego-self is exposed to an environment that is devoid of space and time. If the mind was prepared for such a change, then dissolution of the physical structure would be perfectly natural; such a transition would be associated with relief. Such experiences are not unknown to us. The enormous condensation of impressions which occurs in our dreams or the exaltation of memory associated with a near-death experience, reveals the mind's capacity to deal with this type of transition. This state does not seem to be merely a passive state of expectation. It is rather a state in which the mind catches a glimpse of fresh aspects of reality and prepares itself for adjustment to this new reality. However, this must be a state of great psychic unhingement for a rational mind whose ego-self is strongly attached to the spatial-temporal order. In such a case, this transition is a rather stressful realization of one's wasted opportunities. This stressful experience is a sort of corrective remedy needed for a hardened ego-self to become more sensitive to reality. The mind will have to struggle with its own weaknesses, i.e., making that effort which it did not make during its earthly life. When the effect of its weaknesses has been nullified, it will be fit to continue the onward journey. This is the reason why a painful realization of failure is not everlasting. This sort of experience is meant to clean man's mind of the dross which is a hindrance in his evolutionary progress. When that purpose has been realized, the need for the correction vanishes.[61]

It is these transitory experiences of relief or unhingement by the rational mind in the buffer zone which are referred to in the scriptures and religious literature as "paradise" and "hell." These experiences of "paradise" or "hell" apply to the state before entering into the lowest zones of the Macrocosm, i.e., before the "first heaven." Therefore, "paradise" is not a state to enjoy the rewards for one's previous good deeds. It is rather the starting point of the continuation of the journey. Those in the buffer zone will not be idle, but they will be continually exerting themselves to reach higher states. What is important is that, during that cosmic travel, a person's mind does not lose its individuality. The individuality of every surviving mind is preserved. It is the mind's individuality that may contribute to the enrichment of the Macrocosm.

The terms "heaven" and "hell" are two extremes within a spectrum of experiences that the human mind is exposed to after physical death. It is important to emphasize that these are transitory states of mind; these are not some sort of localities.

In this context, the human life consists of deeds which either reinforce the prominence of the ego-self or contribute to its taming. The deeds prepare one's mind for a future "career." Such a "career" may lead to immortality. Personal immortality, however, is not man's right. It is to be achieved by personal effort. Man, in his natural state, is only a candidate for it.

---

[61] *Islamic Sufism*, The Sirdar Ikbal Ali Shah (Tractus Books, Reno, NV, 2000, p. 198).

***

Physical death is only one of the series of deaths that the human mind experiences. Let's recall that there are several states of heightened consciousness. Each of these states corresponds to a specific layer of the mind. These states form a ladder leading to the final destination. Passing through the various steps of the ladder is referred to as "deaths." This means that man has to experience several "deaths" along his journey. These "deaths" may be experienced in this world, before man's physical death. This is the meaning of the statement that man must "die before he dies." If not, then they are to be experienced afterward.

Each "death" is associated with the release of a particular set of mental or emotional attachments. After "dying" to a particular state, the successful traveler is "resurrected" and only then may attempt to approach the next state. Again, he has to die to be resurrected onto the next and higher state. Each death is followed by "rebirth" or the transformation which results from it. This is like traveling through a succession of islands. Each island represents a certain state. This situation is illustrated by Idries Shah in the following fable:

> Once upon a time there lived an ideal community in a far-off land. Its members had no fears as we now know them. … Although there were none of the stresses and tension which mankind now considers essential to its progress, their lives were richer, because other, better elements replaced these things. Theirs, therefore, was a slightly different mode of existence. We could almost say that our present perceptions are a crude, makeshift version of the real ones which this community possessed.
>
> They had real lives, not semi-lives. …
>
> They had a leader, who discovered that their country was to

become uninhabitable for a period of, shall we say, twenty thousand years. He planned their escape, realizing that their descendants would be able to return home successfully, only after many trials.

He found for them a place of refuge, an island whose features were only roughly similar to those of their original homeland. Because of the differences in climate and situation, the immigrants had to undergo a transformation. This made them more physically and mentally adapted to the new circumstances; coarse perceptions, for instance, were substituted for finer ones, as when the hand of the manual laborer becomes toughened in response to the need of his calling.

In order to reduce the pain which a comparison between the old and new states would bring, they were made to forget the past almost entirely. Only the most shadowy recollection of it remained, yet it was sufficient to be awakened when the time came. …

The system was very complicated, but well arranged. The organs by means of which the people survived on the island were also made the organs of enjoyment, physical and mental. The organs which were constructive in the old homeland were placed in a special form of abeyance, and linked with shadowy memory, in preparation for its eventual activation.

Slowly and painfully the immigrants settled down, adjusting themselves to the local conditions. The resources of the island were such that, coupled with effort and a certain form of guidance, people would be able to escape to a further island, on the way back to their original home. This was the first of a succession of islands upon which gradual acclimatization took place.

The responsibility of this "evolution" was vested in those individuals who could sustain it. These were necessarily only a few, because for the mass of people the effort of keeping both sets of knowledge in their consciousness was virtually

> impossible. One of them seemed to conflict with the other one. Certain specialists guarded the "special science."
>
> This "secret," the method of effecting the transition, was nothing more or less than the knowledge of maritime skills and their applications. The escape needed an instructor, raw materials, people, effort and understanding. Given these, people could learn to swim, and also to build ships.
>
> The people who were originally in charge of the escape made it clear to everyone that a certain preparation was necessary before anyone could learn to swim or even take part in building a ship. …
>
> … The learning and the exercise of this lore depends upon special technique. These together make up a total activity, which cannot be examined piecemeal, … . This activity has an impalpable element, called *baraka*, from which the world 'barque' - a ship- is derived. This word means 'the Subtlety.'[62]

In view of this fable, the buffer zone may be thought of as "a further island," that is the next island on the way back to the "original home." In this fable Idries Shah indicates that the time window prescribed for the "return" is in the range of "twenty thousand years."

***

Life is one and continuous. Man marches always onward to receive ever fresh illuminations from an infinite Reality. The recipient of divine illumination is not merely a passive recipient. Every act of a perfected mind creates a new situation, and thus offers further opportunities of creative unfolding.

---

[62] *The Sufis*, Idries Shah, p. 1-4 (see Note #13).

The Dynamic Cosmos

> For a long time the Universe has been germinating in the very marrow of your bones.
> (*Khaja Hafiz*)

Man may fulfil his evolutionary function by climbing through the various levels of the Macrocosm. In this way he can reproduce the Macrocosm within himself.

The various strata of the Macrocosm are not static; they keep changing, they are alive. They keep changing in accordance with the evolutionary progress of mankind. As mentioned earlier, there is a continuous feedback between the overall state of consciousness of mankind and the composition of the various levels within the Macrocosm. The feedback between the Macrocosm and the physical world operates in accordance with a mechanism known as entanglement. As described by physicists, there is no time delay in this "spooky action at a distance." Each major spiritual breakthrough at the level of ordinary man is immediately marked by a change in the world of symbols and the world of ideas. This sort of entanglement applies to all the layers of cosmic consciousness.

By evolving from his natural state and reaching towards his sublime origins, man actively contributes to the "splendour" of the cosmological structure. This situation is illustrated symbolically in the "Hymn of the Soul" of the *Acts of Thomas*. Here is a version of the Hymn from the tale entitled "The King's Son." The country of Sharq referred to in the tale corresponds to the Macrocosm; man's intellectual and emotional attachments to the physical world (Misr) constitute a fearsome monster, the ego-self:

ONCE in a country where all men were like kings, there lived a family, who were in every way content, and whose surroundings were such that the human tongue cannot describe them in terms of anything which is known to man today. This country of Sharq seemed satisfactory to the young prince Dhat, until one day his parents told him: 'Dearest son of ours, it is the necessary customs of our land for each royal prince, when he attains a certain age, to go forth on a trial. This is in order to fit himself for kingship and so that both in repute and in fact he should have achieved -by watchfulness and effort- a degree of manliness not to be attained in any other way. Thus it has been ordained from the beginning and thus it will be until the end.'

Prince Dhat therefore prepared himself for his journey, and his family provided him with such sustenance they could: a special food which would nourish him during his exile, but which was of small compass though of illimitable quantity.

They also gave him certain other resources, which it is not possible to mention, to guard him, if they were properly used.

He had to travel to a certain country, called Misr, and he had to go in disguise. He was therefore given guides for the journey, and clothes befitting his new condition: clothes which scarcely resembled one royal-born. His task was to bring back from Misr a certain Jewel, which was guarded by a fearsome monster.

When his guides departed, Dhat was alone, but before long he came across someone else who was on a similar mission, and together they were able to keep alive the memory of their sublime origins. But, because of the air and the food of this country, a kind of sleep soon descended upon the pair, and Dhat forgot his mission.

For years he lived in Misr, earning his keep and following a humble vocation, seemingly unaware of what he should be doing.

By means which was familiar to them but unknown to other people, the inhabitants of Sharq came to know of the dire situation of Dhat, and they worked together in such a way as they could, to help release him and to enable him to persevere with his mission. A message was sent by a strange means to the princeling saying: 'Awake! For you are the son of a king, sent on a special undertaking, and to us you must return.'

This message awoke the prince, who found his way to the monster, and by the use of special sounds, caused it to fall into sleep; and he seized the priceless gem which it had been guarding.

Now Dhat followed the sounds of the message which had woken him, changed his garb for that of his own land, and retraced his steps, guided by the Sound, to the country of Sharq.

In a surprising short time, Dhat again beheld his ancient robes, and the country of his fathers, and reached his home. This time, however, through his experiences, he was able to see that it was somewhat of greater splendour than ever before…[63]

Let's recall that the *creative mind*, the *sublime mind*, and the *supracognitive mind* do not exist naturally. They did not appear in those forms in the original arc of descent. These minds ("perfected souls") are a result of man's special undertaking. In their original forms these minds ("souls") were not endowed with earthly experiences. Instead, they were arranged according to their initial

---

[63] *Tales of the Dervishes*, Idries Shah, p. 217 (See Note #27).

predispositions. Here is a story which explains these different predispositions:

> It is recorded in the traditions of the Lovers of Truth, that when the souls were created, before the bodies, they were asked what they wanted as a means of traveling in this world.
>
> There were four parties among them. The first desired to travel on foot, as the safest method. The second desired horses, for this would mean less work for them. The third wished to travel on the wind, to overcome limitations. The fourth chose light, by which they could understand as well as move.
>
> These four groups still exist, and all people still abide by one of these characteristics. …[64]

In the above story, the first group is driven by the self-faculty; the second group is under the influence of emotions; the third group is mostly directed by the intellect faculty. We may notice that these three groups' preferences correspond to those whose destination in their present life is the buffer zone; their predisposition is towards the re-arrangement of their *rational mind*. The fourth group are those whose preferences are focussed on the Macrocosm. Their potential corresponds to experiencing zones of the invisible worlds during this life.

These various predispositions do not determine the final states. Instead, they correspond to assigned potentialities in accordance with the cosmic needs. An individual "soul" may or may not fulfil its potentiality. In some cases, during the journey its potentiality

---

[64] "The Four Types" included in *Seeker After Truth*, Idries Shah (The Octagon Press, London, 1982, p. 14).

may be upgraded or degraded. This is indicated by the saying recorded in Attar's *Recital of the Saints*:

> The people of the world have a fixed destiny. But the spiritually developed receive what is not in their destiny.[65]

There is a certain minimum of enrichment of the Macrocosm needed to warrant the sustenance of the physical world. Let's recall that after the disintegration of what appears to be man's personality, a person's mind will preserve its individuality. It is this individuality that will contribute to the increase of the richness of the Macrocosm. The minimum of enrichment constitutes a certain critical mass of human "souls" needed to keep flowing into the various layers of the Macrocosm. This flow of enrichment is needed to balance out mankind's destructive impact on the planet's natural zone of consciousness. It is in this manner that the physical world is kept alive.

Let's recall that during his first ascent, Mohammed passed through the "seventh heaven" and arrived within the symbolic distance of "two bows" from the Absolute. At that level, there were no other "souls." This was symbolically indicated by the fact that even the angel Gabriel (one of the highest of the original cosmic minds) could not reach that state of consciousness:

> O Mohammed, if I take one more step, it will burn me;
> Do thou leave me, henceforth advance alone: this is my limit.
> (*Mathnawi, Book I, 1066-7*)

---

[65] Quoted by Idries Shah in *Learning How to Learn* (The Octagon Press, London, 1981, p. 23).

Yet, the distance of "two bows" is still infinitely large in comparison to "We are nearer to him than his jugular vein." At that time, Mohammed was not ready yet to be annihilated within the Absolute; he still had "a certain darkness"[66] in his nature:

> Like you, I was dark in my nature: the Sun's revelation gave me such a light as this. I have a certain darkness in comparison with the spiritual suns, but I have light for the darknesses of human souls. I am less bright than the Sun in order that thou mayst be able to bear my beams, for thou art not the man for a man who can bear the most radiant Sun.
> (*Mathnawi, Book I, 3660-2*)

Mohammed could not enter the "eighth heaven." He had to return to the ordinary world to complete his mission. Only when he died, could he start his final ascent. At one point in his final ascent, his sublime part gradually increased to such an extent that no traces of earthly attachments remained. After reaching the highest state, the Macrocosm became "somewhere of greater splendour than ever before." This was a major milestone of human evolution. The first perfected earthly soul arrived at its ultimate destination.

It is this sort of human experience that changes the composition of the Macrocosm. This change indicates that there is a qualitative hysteresis between the descending arc and the ascending arc. The measure of this spiritual hysteresis is expressed as the symbolic difference between "two bows" which mark Mohammed's first ascent and "We are nearer to him than his jugular vein" which

---

[66] This "darkness," or earthly attachments, is indicated symbolically in the description of the Night Journey by physical effects (e.g., "A pot of water which had been overturned by the flight and spilled was still not empty when the Prophet returned." – see the story "The Sultan Who Became an Exile").

describes the state of his second ascent. This means, that although both the descent and the ascent lead through the same strata of the Macrocosm, each ascending step is of higher quality than the corresponding descending step. In is in this sense that the perfected (returning) souls contribute to the enrichment of the Cosmos. They are the second cycle of souls, while the original cosmic souls constitute the first cycle. Although both types are in the closest vicinity to the Absolute, there is a vast gulf between them.

In Attar's poem, the critical flow was symbolically presented as "thirty birds." At the end of their journey, the thirty birds formed an ascending triplicity that matched the original triplicity. It is this form of triplicity that needs to be achieved within each layer of the Macrocosm in order to preserve not only the Universe but the entire Cosmos. The spiritual quality of each layer has to be kept sustained by populating it with perfected earthly souls.

The transfer of this sort of critical mass to each of the macrocosmic layers provides a continuous feedback to the cosmic infrastructure. It is also this flow that controls the timing mechanism that triggers the major milestones of the process of creative synthesis.

To ensure that the overall process continues, a critical mass of perfected human "souls" needs to flow through the various strata of the Macrocosm. These various strata (or spheres) of the Macrocosm may be compared to the set of strings of a musical instrument. The arrival of perfected souls makes them oscillate at "frequencies" specific to their locations within the Macrocosm. Together, these oscillations generate "special sounds" which are often referred to as the "music of the spheres." It may be said that men may be ennobled (gain immortality) by tuning their minds into resonance with this music that "we cannot hear it." Shakespeare's Lorenzo comments on it in the following quote:

There's not the smallest orb which thou behold'st
But in his motion like an angel sings,
Still quiring to the young-ey'd cherubins;
Such harmony is in immortal souls;
But whilst this muddy vesture of decay
Doth grossly close it in, we cannot hear it.
(*The Merchant of Venice, V.1*)

Every generation of humanity has to "produce" its minimal critical mass contribution to the Macrocosm. It is this flow that contributes to the dynamics of the Cosmos. The presence of "perfected" souls within the Macrocosm has to keep increasing in order to overcome the entropy of human existence on the planet.

As indicated previously, there are three main levels of consciousness within the Macrocosm: the Realm, the world of ideas, and the world of symbols. On each of these levels there is a need for a gradual increase of the presence of perfected human souls. At one point, men will gradually disappear and become extinct; the planet will cease to provide its "raw material." At that future time, there will have to be enough perfected human souls on all levels of the Macrocosm to warrant the continuation of the cosmic evolution. At that future time, a New Cosmos will be formed. For this New Cosmos to appear, all the levels of the Macrocosm will have to be populated with perfected earthly souls.

***

There is a specific pattern of distribution on every level of the Macrocosm that is needed to be completed during the current evolutionary cycle. It was this distribution pattern that dictated the originally assigned predispositions of the cosmic souls. Hints of this "mystical" pattern have been given in the past, some more overt than others. Here is a symbolic reference to such a distribution ("the first mystery" refers to the structure of the current Cosmos):

> I say to you, there will be found one in a thousand, two in ten thousand, for the completion of the mystery of the first mystery.
> (*The Gospel of Thomas,* Note *23*)

This distribution keeps changing in accordance with mankind's evolutionary progress. It remains unknown to the general public. However, when there is a major adjustment, hints about the change are discreetly disclosed. The first insight into the distribution pattern was provided by the account of the Night Journey. Afterwards, the pattern has been symbolically reflected in the earthly administrative structure of those who have been actively involved in the process. The adjustments to that administrative structure are indicative of the changes within the Macrocosm.

A modern version of that administrative structure was established at the time when Mohammed concluded his final ascent. In the context of the physical world, his final ascent lasted one thousand years. (This is why it is said that "It is only once in a thousand years that this secret is seen by man. When he sees it, he is changed.") After that period, his link to the physical world ceased to exist:

A totally wise man would cease to exist in the ordinary sense.[67]

This disconnection marked the beginning of the second spiritual millennium of the modern world. It was then that the overall "mystical" structure was adjusted accordingly. This took place at the end of the 16th century. At that time, another individual was invested with the responsibility of overseeing the process on the Earth. This particular individual, Ahmed Farugi of Sirhind, is known as the "Regenerator of the second spiritual millennium." At that time, the structure had a pyramid-like multi-level shape. The various levels were reflections of the strata of the Macrocosm. The population of the levels was decreasing while moving from the lower to the top levels. At the top was the Axis, the Complete Man. On the second level, there were four Deputies. The next levels were populated by Ennobled Men of the various ranks. There were seven, three, and forty Ennobled Men, respectively. Here is a symbolic representation of the top five levels of that structure:[68]

Δ

Δ Δ Δ Δ

Δ Δ Δ Δ Δ Δ Δ

Δ Δ Δ

Δ Δ Δ Δ Δ Δ Δ Δ Δ Δ Δ Δ Δ Δ Δ Δ Δ Δ Δ Δ

Δ Δ Δ Δ Δ Δ Δ Δ Δ Δ Δ Δ Δ Δ Δ Δ Δ Δ Δ Δ

---

[67] "Meditations of Rumi" in *Caravan of Dreams*, Idries Shah (The Octagon Press, London, 1968, p. 79).

[68] *Revealed Grace*, Arthur F. Buehler (Fons Vitae, Louisville, KY, 2011, p. 272).

The most recent change was implemented in the late 1960s. It was then that another adjustment was made. Another individual was charged with the function of overseeing and directing the evolutionary process. It was then that the responsibility, which until then was divided among the four Deputies, once again converged on one man. The new structure took on the following shape:[69]

Δ
Δ Δ Δ Δ
Δ Δ Δ Δ Δ Δ Δ
Δ Δ Δ Δ Δ
Δ Δ Δ Δ Δ Δ Δ Δ Δ Δ Δ Δ Δ Δ Δ Δ Δ Δ Δ Δ
Δ Δ Δ Δ Δ Δ Δ Δ Δ Δ Δ Δ Δ Δ Δ Δ Δ Δ Δ Δ
Δ Δ Δ Δ Δ Δ Δ Δ Δ Δ Δ Δ Δ Δ Δ Δ Δ Δ Δ Δ
Δ Δ Δ Δ Δ Δ Δ Δ Δ Δ

Although the top three layers remained unchanged, there was a change within the lower parts of this structure. The numbers of Ennobled Men on the fourth and fifth levels increased from three to five and from forty to seventy, respectively. This change within the earthly structure symbolically indicates the progress achieved over the last several centuries. It is a reflection of the enrichment of the corresponding layers within the Macrocosm. It is this change that allows to gauge the status of the entire process with respect to the original plan.

This "tiny" enrichment affects the entire human race; it is needed to sustain the physical world. It is a measure of human progress. If there was no such progress, the physical world would collapse.

---

[69] *Journeys with a Sufi Master*, H.B.M. Dervish (The Octagon Press, London, 1982, p. 153).

Rafael Lefort, a writer who recorded his Asian travels in *The Teachers of Gurdjieff*, refers to this situation in his book:

> The people in those centers are concerned with the destiny of the world. … They are no ordinary men, let alone monks. They know neither rest nor even satisfaction, for they have to make up for the shortcoming of humanity. They are the Real people who have experienced being and non-being and have long ago entered a stage of evolution when neither state means anything to them.[70]

This enrichment also controls the "trigger" for the major milestones of human evolution. It is this enrichment, or lack of it, that will delay or accelerate the "end of time."

The above two diagrams show the top of the cross sections of the structure at two stages of the evolutionary process. There are many more levels and empty "spots" which need to be filled in. The various positions of the empty spots correspond to the various potentialities of men's minds. In *As You Like It*, Shakespeare's young protagonist, Orlando, alludes to such "empty spots":

> Only in the world I fill up a place, which may be better supplied, when I have made it empty.
> (*As You Like It, I.2*)

---

[70] *The Teachers of Gurdjieff*, Rafael Lefort (Victor Gollancz Ltd., London, 1966, p. 96).

The evolution of man is to be continued till the completion of the macrocosmic structure. At that future time, the physical Universe will cease to exist and all of humanity will be transmuted into "elements" of the Macrocosm. A new "table of elements" will be formed. This future structure will form a New Cosmos. The New Cosmos will consist only of the enriched Macrocosm. Only then the function of mankind will be fully completed.

Now we may fully comprehend what the ultimate goal of mankind is. By developing higher levels of consciousness, mankind is being transmuted into new "elements" that are needed for the creation of a New Cosmos. It is the creation of a future New Cosmos that Rumi's previously quoted statement in the Chapter on The Human Mind refers to:

> There are a hundred thousand more marvellous states ahead of him.

The creation of a New Cosmos is the ultimate purpose of human existence.

***

The Perfect Men's prime function is to ensure that progress continues. Without progress, the timing mechanism would be activated and humanity would be eliminated prior to fulfilling its evolutionary function. Rumi is quite clear on this point:

> If the Perfect Man would vanish, destiny would come upon us
> and the entire world would cease to exist.
> (*Mathnawi, Book I, 99*)

Shakespeare echoes Rumi's view:

> ... the times should cease,
> And threescore year would make the world away.
> (*Sonnet 11*)

As the evolutionary process is advancing, new developmental techniques are being made available to the mystics who are in charge of the human evolution. This is why there are different techniques and developmental methodologies that were implemented in various historical times. For example, as a result of Mohammed's entering into the "eighth heaven," the mystics gained access to a compressed form of the spectrum of the modes of consciousness that are available in the Macrocosm. This allowed for the introduction of a new developmental methodology. By using this methodology, a person may be simultaneously exposed to the entire spectrum of the field of universal consciousness that is available in the Macrocosm. This new methodology was introduced by Ahmad Sirhindi at the end of the 16th century. Sirhindi used the term "the world of directive energy" to describe the compressed form of the spectrum of consciousness.[71] Through exposure to directive energy, it is possible to activate simultaneously several

---

[71] *Islamic Sufism*, The Sirdar Ikbal Ali Shah, p. 97 (see Note #61).

subtle faculties. This allows for a greatly accelerated development of the inner structure of the mind. Of course, such an exposure to the spectrum of consciousness required new advanced techniques that were not available in the past. This means that the developmental methodology had to be modified. It had to be customized in accordance with a person's natural characteristics. The applied spectrum of consciousness has to match those levels of the subtle faculties that are naturally stronger within a particular person's mind. Otherwise its application would be harmful. With the introduction of this new methodology, the previously used approaches and techniques became obsolete.

Ahmad Sirhindi illustrates symbolically an outline of this new methodology in the story "The Lame Man and the Blind Man." The lame man and the blind man represent the ordinary faculties of heart and intellect, respectively. None of these faculties on its own is capable of moving forward. In their natural states, these faculties are deficient; they cannot bring man to his ultimate destination ("to reach the king's banquet"). The story explains how, despite these deficiencies, progress can be made. Namely, in the presence of an observer ("a third man"), the latent faculties may be activated and in this way the ordinary faculties can overcome their limitations. The observer's instruction is a symbolic illustration of an exposure to a customized spectrum of directive energy:

> A LAME man walked into a Serai ('Inn') one day, and sat down beside a figure already seated there. 'I shall never be able to reach the king's banquet,' he sighed, 'because due to my infirmity, I am unable to move fast enough.'
>
> The other man raised his head, 'I, too, have been invited,' he said, 'but my plight is worse than yours. I am blind, and cannot see the road, although I have also been invited.'
>
> A third man who heard them talking said: 'But, if you only

> realized it, you two have between you the means to reach your destination. The blind man can walk, with the lame one on his back. You can use the feet of the blind man and the eyes of the lame to direct you.'
>
> Thus the two were able to reach the end of the road, where the feast awaited them. ...[72]

The interesting thing is that this latest methodology was nearly immediately transferred from India to late 16th century England. It was part of the preparation for the commencement of the next phase of the evolutionary process on the planet. For the first time in human history, the new evolutionary phase was to be initiated entirely within the secular milieu of western society. This new methodology was disclosed within Shakespeare's writings.

Shakespeare used his narrative composed of 37 plays to describe 37 episodes of the evolutionary process that led to the formation of western civilization. Shakespeare treats historical events as manifestations of the state of the mind of a select group of people who are representative of a given geographical area at that particular time. As indicated earlier, it is the state of the human mind that drives history and the welfare of societies. In Shakespeare's allegorical presentation, his characters are a symbolic representation of various faculties of the mind. Some of these faculties are ordinary; some are extraordinary, and some others are still in their latent state. It is such a composite state of the mind that determines what is possible and what is impossible; it defines its evolutionary potential and dictates the sequence of events. The narrative starts with the Trojan War, moves to ancient Greece and pre-Roman Britain, then goes to Rome, continues through the Middle Ages and concludes with the appearance of the European

---

[72] *Tales of the Dervishes*, Idries Shah, p. 209 (see Note #27).

Renaissance.[73] Shakespeare introduced a unique illustration of the application of the new methodology of the activation of the inner faculties. Namely, the evolutionary progress is indicated by a number of couples who get married simultaneously ("all the four, or the three, or the two, or one of the four.") In this way he symbolically illustrates the implementation of the developmental methodology in various geographical areas. It was this new methodology that led to the formation of modern western society.

One would have to be familiar with the methodology of the activation of the inner layers of the mind to recognize the "secret" that is hidden in Shakespeare's plays. Just as in the case of Maxwell's equations, for a person who is not familiar with this symbolic illustration, such an interpretation of the plays neither makes sense nor has any importance.

***

It is the early spring of 1590. Little Athens, a charming town in northern Italy, is getting ready for festivities organized by Duke Gonzaga, the ruler of the Dukedom of Sabbioneta. Little Athens is the capital of the Dukedom. The town is located seven leagues southwest of Mantua. In the centre of the town there is a Column of Pallas-Athena. Duke Gonzaga erected the column to indicate that Athena, the Greek goddess of wisdom and the Chief Muse, is the patroness of his city.

There is a lot of excitement in the town. As part of the festivities there will be performances in the recently built theatre, Teatro all'Antica. The new theatre is not an ordinary theatre. It is the first free-standing, roofed and purpose-built

---

[73] *Shakespeare's Elephant in Darkest England*, Wes Jamroz, p. 200 (see Note #35).

theater in the modern world.

At the city's western gate, called the Duke's Oak, there is a group of men. (The gate is called the Duke's Oak because it leads onto the Duke's hunting ground in an oak forest.) They are actors who have been invited by the Duke to stage their performance in the new theatre. The actors just came back from the oak forest, where they were rehearsing their performance. According to a local legend, the forest is haunted by fairies. There are many fascinating stories about the fairies told by the locals. Recently, a couple of the Duke's courtiers experienced weird encounters with the supposed fairies in the oak forest. Since then, the city has been buzzing with all sorts of gossips and rumours.

The Duke asked the actors to stage a play which would help his guests and his courtiers grasp the idea of human evolution. Particularly, the Duke wanted the actors to address the concept of the invisible worlds and the role of the Perfect Man.

After picking up their costumes and props from a small room within the gate, the actors turned left and walked along a street leading to the Temple.

While walking the streets of Little Athens one may discover how unique the city's design is. The entire city was constructed in the mannerist architectural style. It was designed as a fortress and a number of mannerist devices were used as means to confuse potential invaders. The supposed symmetrical layout of the city is intentionally misleading. The location of the two main squares is decentralized in relation to the geographical centre line of the city. A false perspective was used by gradually decreasing the width of the streets. This made the streets appear longer than they really are. All of these features turned the city into a sort of labyrinth.

The Temple is a powerful octagonal redbrick structure with reinforced and protruding corners. Inside, there are eight radial niches surmounted by a gallery. A couple of years earlier, the Duke had demolished a church and a priory that had been there before. In their place, he built the octagonal Temple. It is in this Temple that two of the Duke's courtiers were married a few hours earlier.

The actors pass by the Temple, then turn right and arrive in the front of the Duke's palace. The Duke and his guests have just finished supper and are getting ready to walk to the theatre, which is located just one block south of the palace. Among the Duke's guests are nobility and intelligentsia from Italy and other western European countries. During the past few days, the guests had a chance to admire the Duke's rich art collections and attend scholarly lectures that he sponsored.

The new theatre building has three separate entrances, one reserved for the Duke and his family, one for the courtiers and the Duke's guests, and one for the actors. Inside the theatre there is a permanent stage design representing a typical Italian city. There are a few novelties in the overall design of the theatre. Namely, there is the double foyer, separate for men and women, the dressing rooms, and the stage door.

In their play, the actors use a couple of lovers to represent the evolutionary state of mankind. At the beginning of the play, the lovers get into unexpected difficulties. The play indicates that these difficulties are similar to those encountered by Pyramus and Thisbe, the famous lovers from the ancient city of Babylon at the beginning of the second millennium BC. These difficulties are remnants of the evolutionary disruption that occurred in antiquity. It seemed that since then all lovers all over the world had been faced with the same situation, including Romeo and Juliet of the famous story that originated in 14th century Verona.

The actors' play is an allegorical illustration of a cross section of the human mind. The word of ideas is represented by the moon. The world of symbols is represented by the Kingdom of Fairies. The lovers belong to the ordinary world. The play explains that the lovers cannot be united because there is a certain disharmony within the Fairyland. It is this disharmony projected onto the ordinary world that has affected all lovers throughout history. This disharmony is projected onto the ordinary world by the symbols of "hungry lion" and "wall." The "hungry lion" represents sensual attractions which interfere with the lovers' relationships. The "wall" is the inadequacy of the entire environment around the lovers, which is manifested as hatred between the lovers' parents. Men of the ordinary world are incapable of solving this situation. It does not matter what they try, their situation cannot be changed. The situation could only be mended by a wise man capable of entering into the world of symbols (Fairyland) and fixing the problem there. In order to make this journey, this man had to be transmuted in a certain way. The actors symbolically present this as a transformation of the man into an "ass." (An "ass" is equivalent to a "fool," i.e., a special fool: "the fool doth think he is wise, but the wise man knows himself to be a fool.") There was, however, another requirement. The trip to the Fairyland could only take place at a specific time dictated by the world of ideas. This time is symbolically marked by the full moon.

The actors stage their play in five episodes, which may be entitled "The Wall," "The Bloody Mantle," "The Lion," "The Moon," and "The Lovers' Death." These episodes are arranged in reverse chronological order, starting with the last, i.e., the most recent event ("The Wall"). In this way they indicate to their audience that their story is an allegory set-up in imaginary time.[74] The initial episode, "The Lovers' Death," refers to the hopeless situation of all "lovers" since antiquity. The "Moon" marks the time when it was possible for a wise man to enter the Fairyland. "The Bloody Mantle" is a sign that the hungry "Lion" has been subdued. The removal of the "Wall" in the last episode is a sign that the situation within the Fairyland has been fixed. It was that "fix" that solved the lovers' problem. At

---

[74] The interpretation of "Pyramus and Thisbe" was extracted by the author from *Shakespeare's Elephant in Darkness England*, W. Jamroz, p. 328 (see Note #35).

the end, the lovers are happily united.
By staging their play, the actors attempt to explain to the young lovers their weird experiences in the oak forest. In their symbolic language, the actors are explaining why the lovers could finally be happily married. They could be united because a wise man had been capable of entering into the world of fairies and fix the problem there.
The actors present their play in the mannerist style. They apply a "false perspective" to ordinary perceptions by skillfully caricaturizing themselves and their characters. They make their story incomprehensible to those who are mostly driven by intellectual responses and emotional reactions. As long as man is driven by such inferior impulses, he is incapable of grasping the symbolic meaning of the play. In this way the actors mock nature by perfectly imitating her.
The conclusion of the actors' story summarizes quite adequately the state of the evolutionary process that was implemented in Western Europe at the end of the 16th century.

We can recognize that the entire plot of the play is a reference to the evolutionary milestone that was accomplished at the end of the 16th century. For the first time in the world literature, a man was able to enter the "Fairyland" and fix a problem within the world of symbols. This was possible because, at that time, another man managed to be fully absorbed within the Absolute. In the ordinary word, this specific refinement of the human mind was manifested as the appearance of the European Renaissance.

The Duke's guests and courtiers, however, did not grasp the meaning of the play. Instead, they were roaring with laughter watching the actors as they were supposedly struggling with their production. The spectators were too arrogant to pause and take another look at the situation that was unfolding in front of them. If they would have paused for a moment, they would have recognized that the actors' story contained "all that you are like to know." Fortunately, the play was written down; it has been passed down through generations under the title *A Midsummer Night's Dream.* Since then, it has been replayed many times over the period of the last four hundred years or so. Therefore, many generations of spectators have had a chance to learn all that they needed to know about how to contribute to the true enrichment of the world.

Why Are We Here?

> The purpose of the exercise of the science of knowledge is to gain an eternally durable existence.
> (*Al Ghazali*)

Based on the presented model of cosmic consciousness, the following observations may be drawn to summarize the modus operandi of the Universe, the purpose of life and the role of humanity and its future. These observations contain enough information to address the "childish questions" mentioned in the first Chapter. One does not need to accept the existence of a divine intervention or agree with the mystical methodology in order to use them. These observations are just descriptions of the operation of the field of universal consciousness:

- Consciousness is a form of energy.
- The Cosmos is a gradient of consciousness and the Universe occupies the lowest level within the cosmic consciousness.
- In his natural state, man is equipped with physical faculties that are limited to perceiving the physical world.
- Other types of faculties are needed to perceive and operate within heightened levels of consciousness.

- The human mind contains a set of inner faculties that are sufficient for perceiving all the levels of consciousness. However, these more subtle faculties are in their latent states.

- By activating his latent faculties, man is capable to reach towards the highest levels of consciousness.

- From a particular level of consciousness, only the immediately next level is partially perceptible; higher levels remain indiscernible.

- A particular level of consciousness may only be fully developed while struggling towards the next higher one.

- At each level of consciousness, there are different laws concerning space, time, and existence.

- An act of active observation may be performed only on objects that belong to levels which are less subtle than the consciousness of the observer.

- Man's efforts toward the development of higher consciousness constitute a "deliberate evolution."

- The purpose of human life is to evolve to higher levels of consciousness.

These observations provide indications about the place of the physical world within the Cosmos. As such they may be useful in guiding scientists in their attempt at developing an adequate model of the Universe.

These observations may prove to be more effective than the so-called anthropic principle.[75] As a matter of fact they may very well replace it. The anthropic principle has been constructed by the scientific community as a method for addressing such questions as:

> Why are the conditions on the Earth just right for the existence of conscious beings such as humans?

According to the anthropic principle, the answer would be:

> If the conditions were not just right, then we would not be here to ask this question.

It is quite puzzling to see that physicists and cosmologists would ever consider the twisted logic of the anthropic principle to be of any use at all. As pointed out by Roger Penrose, "the anthropic principle tends to be invoked by theorists whenever they do not have a good enough theory to explain the observed facts."[76]

As the above observations indicate, it is impossible from the level of ordinary consciousness to grasp the laws operating within the higher levels of consciousness. This impossibility determines the limits of science. The scientific enterprise takes place within the field of operation of the intellect faculty. In accordance with the overall cosmic structure, such undertaking is limited to understanding the physical world, i.e., the lowest zone of the field of consciousness. Such an approach, therefore, is inefficient for reaching towards higher zones of consciousness. The intellect is too coarse a device to grasp the intrinsicalities of the Macrocosm. In other words, scientific activities will never succeed in the formulation of the "mind of God." However, by striving (unintentionally) towards the world of symbols, scientists

---

[75] The anthropic principle is a philosophical assumption that humans play a privileged role in the self-consistent Universe. This assumption excludes the existence of any form of higher consciousness.

[76] *The Emperor's New Mind*, Roger Penrose, p. 433 (see Note #6).

continually improve the mechanical model of the physical world. In this manner they contribute to the development of technologies that are essential for maintaining the presence of humanity on the planet.

As science discovers new laws, it also introduces new terms, expressions, and concepts. The new technical and scientific vocabulary allows for a more adequate description of the invisible worlds. It is in this indirect manner that science contributes to a better understanding of possibilities associated with heightened states of consciousness.

By taking advantage of the mystical data, the scientists may be able to navigate more efficiently through a number of challenges that humanity is presently faced with. For example, the available mystical data may help in making choices when selecting the most promising approach: Is it more promising to work towards development of the theory of everything or would it be more beneficial for humanity to first research the link between consciousness and dark matter? Is it more valuable to search for inhabitable exoplanets or would it be more beneficial to investigate the effect of long-term presence in space on human cognition? Is it more beneficial for humanity to invest resources in looking for traces of life in the Universe or would it be better to work towards preserving humanity by investigating the effect of the natural environment on human consciousness?

There is a historical precedent of this sort of relationship between different approaches, such as science and mysticism. It concerned philosophy and science. It has been noticed that up to the 18th century, the philosophers were regarded as the leading minds of the world. They considered the whole of human knowledge to be their exclusive field. At one point, however, the philosophers ignored one critical element in their inquiries: experiments. They regarded any form of experimental data as inferior to the purely speculative thinking which they employed as the prime method for drawing conclusions. However, trying to answer the question "why are we here" through the employment of logic and intellect is as impossible as attempting to conceive an idea of the existence of numerous galaxies without a telescope and astrophysical observations. This is why, at one point, the philosophers lost grasp of the advances that were taking place in science. These advances were stimulated by a huge body of experimental data coming from observations and mathematical modeling, activities that were ignored by the philosophers. Consequently, as stated by Ludwig Wittgenstein, one of the most influential philosophers of the 20th century, the scope of their inquiries was reduced so much that "the sole remaining task for philosophy is the analysis of language."[77] It seems that now theoretical physicists are finding themselves in a similar situation as the 18th century philosophers. As they have nearly reached the limits of energies available in their laboratories, their contribution is gradually being reduced to generating new terms and abstract concepts. If they keep ignoring the vast experimental data provided by the mystics, theoretical physicists will soon cut themselves off from a pool of relevant information. Their future activity will be greatly reduced and might become equivalent to gamesters playing with multidimensional models and geometrical objects, a fascinating undertaking but with no developmental significance.

In the meantime, humanity cannot wait for science to determine what is in the "mind of God." Regardless of science's successes or failures, mankind is to discharge its function as an active participant in the process of creation.

---

[77] Quoted by Stephen Hawking in *A Brief History of Time*, p. 174 (see Note #5).

TROUBADOUR PUBLICATIONS

A Journey with Omar Khayaam, W. Jamroz (2018)

Shakespeare's Elephant in Darkest England, W. Jamroz (2016)

El elefante de Shakespeare, W. Jamroz (2016) (*in Spanish*)

Shakespeare's Sequel to Rumi's Teaching, W. Jamroz (2015)

Shakespeare's Sonnets or How heavy do I journey on the way, W. Jamroz (2014)

Shakespeare for the Seeker, Volume 4, W. Jamroz (2013)

Shakespeare for the Seeker, Volume 3, W. Jamroz (2013)

Shakespeare for the Seeker, Volume 2, W. Jamroz (2013)

Shakespeare for the Seeker, Volume 1, W. Jamroz (2012)

Printed in Great Britain
by Amazon

58351158R00108